LEARNING FROM OUR BUILDINGS
A State-of-the-Practice Summary of Post-Occupancy Evaluation

Federal Facilities Council Technical Report No. 145

NATIONAL ACADEMY PRESS
Washington, D.C.

NATIONAL ACADEMY PRESS   2101 Constitution Avenue, N.W.   Washington, DC 20418

NOTICE

The Federal Facilities Council (FFC) is a continuing activity of the Board on Infrastructure and the Constructed Environment of the National Research Council (NRC). The purpose of the FFC is to promote continuing cooperation among the sponsoring federal agencies and between the agencies and other elements of the building community in order to advance building science and technology—particularly with regard to the design, construction, acquisition, evaluation, and operation of federal facilities. The following agencies sponsor the FFC:

Department of the Air Force, Office of the Civil Engineer
Department of the Air Force, Air National Guard
Department of the Army, Assistant Chief of Staff for Installation Management
Department of Defense, Federal Facilities Directorate
Department of Energy
Department of the Interior, Office of Managing Risk and Public Safety
Department of the Navy, Naval Facilities Engineering Command
Department of State, Office of Overseas Buildings Operations
Department of Transportation, U.S. Coast Guard
Department of Transportation, Federal Aviation Administration
Department of Veterans Affairs, Office of Facilities Management
Food and Drug Administration
General Services Administration, Public Buildings Service
Indian Health Service
International Broadcasting Bureau
National Aeronautics and Space Administration, Facilities Engineering Division
National Institute of Standards and Technology, Building and Fire Research Laboratory
National Institutes of Health
National Science Foundation
Smithsonian Institution, Facilities Engineering and Operations
U.S. Postal Service, Engineering Division

As part of its activities, the FFC periodically publishes reports that have been prepared by committees of government employees. Because these committees are not appointed by the NRC, they do not make recommendations, and their reports are considered FFC publications rather than NRC publications.

For additional information on the FFC program and its reports, visit the Web site at <http://www4.nationalacademies.org/cets/ffc.nsf> or write to Director, Federal Facilities Council, 2101 Constitution Avenue, N.W., HA-274, Washington, DC 20418 or call 202-334-3374.

Printed in the United States of America

2001

FEDERAL FACILITIES COUNCIL

Chair

Henry J. Hatch, U.S. Army (Retired)

Vice Chair

William Brubaker, Director, Facilities Engineering and Operations, Smithsonian Institution

Members

Walter Borys, Operations and Maintenance Division, International Broadcasting Bureau
John Bower, MILCON Program Manager, U.S. Air Force
Peter Chang, Division of Civil and Mechanical Systems, National Science Foundation
Tony Clifford, Director, Division of Engineering Services, National Institutes of Health
Jose Cuzmé, Director, Division of Facilities Planning and Construction, Indian Health Service
David Eakin, Chief Engineer, Office of the Chief Architect, Public Buildings Service, General Services Administration
James Hill, Deputy Director, Building and Fire Research Laboratory, National Institute of Standards and Technology
John Irby, Director, Federal Facilities Directorate, U.S. Department of Defense
L. Michael Kaas, Director, Office of Managing Risk and Public Safety, U.S. Department of the Interior
Joe McCarty, Engineering Team Leader, U.S. Army Corps of Engineers
William Miner, Acting Director, Building Design and Engineering, Office of Overseas Buildings Operations, U.S. Department of State
William Morrison, Chief, Structures Branch, Facilities Division, Air National Guard
Get Moy, Chief Engineer and Director, Planning and Engineering Support, Naval Facilities Engineering Command, U.S. Navy
Robert Neary, Jr., Associate Facilities Management Officer, Office of Facilities Management, U.S. Department of Veterans Affairs
Juaida Norell, Airways Support Division, Federal Aviation Administration
Wade Raines, Maintenance and Policies Programs, Engineering Division, U.S. Postal Service
James Rispoli, Director, Engineering and Construction Management Office, U.S. Department of Energy
William Stamper, Senior Program Manager, Facilities Engineering Division, National Aeronautics and Space Administration
Stan Walker, Division Chief, Shore Facilities Capital Asset Management, U.S. Coast Guard

Staff

Richard Little, Director, Board on Infrastructure and the Constructed Environment (BICE)
Lynda Stanley, Director, Federal Facilities Council
Michael Cohn, Program Officer, BICE
Kimberly Goldberg, Administrative Associate, BICE
Nicole Longshore, Project Assistant, BICE

Preface

At the most fundamental level, the purpose of a building is to provide shelter for activities that could not be carried out as effectively, or carried out at all, in the natural environment. Buildings are designed and constructed to (1) protect people and equipment from elements such as wind, rain, snow, and heat; (2) provide interior space whose configuration, furnishings, and environment (temperature, humidity, noise, light, air quality, materials) are suited to the activities that take place within; and (3) provide the infrastructure—water, electricity, waste disposal systems, fire suppression—necessary to carry out activities in a safe environment.

Today, people and organizations have even higher expectations for buildings. Owners expect that their investments will result in buildings that support their business lines or missions by enhancing worker productivity, profits, and image; that are sustainable, accessible, adaptable to new uses, energy efficient, and cost-effective to build and to maintain; and that meet the needs of their clients. Users expect that buildings will be functional, comfortable, and safe and will not impair their health. A building's performance is its capacity to meet any or all of these expectations.

Post-occupancy evaluation (POE) is a process for evaluating a building's performance once it is occupied. It is based on the idea that finding out about users' needs by systematically assessing human response to buildings and other designed spaces is a legitimate aim of building research. Early efforts at POE focused on housing needs of disadvantaged groups to improve environmental quality in government-subsidized housing. The process was later applied to other government facilities such as military housing, hospitals, prisons, and courthouses. POE began to be used for office buildings and other commercial real estate in the mid-1980s and continues to be used for a variety of facility types today.

As POE has been applied to a larger range of building types and as expectations for buildings have evolved, POE has come to mean any and all activities that originate out of an interest in learning how a building performs once it is built, including whether and how well it has met expectations and how satisfied building users are with the environment that has been created. Although POEs are still focused on determining user comfort and satisfaction, organizations are attempting to find ways to use the information gathered to support more informed decision-making about space and building investments during the programming, design, construction, and operation phases of a facility's life cycle. To do this, organizations need to establish design criteria, databases or other methods for compiling lessons from POEs and for disseminating those lessons throughout the organization, from senior executives to midlevel managers, project managers, consultants, and clients.

The federal government is the United States' largest owner of facilities, with approximately 500,000 facilities worldwide. Federal agencies that own, use, or provide facilities have a significant interest in optimizing their performance. The General Services Administration, U.S. Army Corps of Engineers, Naval Facilities Engineering Command, U.S. Postal Service, State Department, and Administrative Office of the U.S. Courts have been leaders in the development and practice of POEs. They and other federal agencies are trying to find ways to share information about effective processes for conducting POEs, to capture and disseminate lessons learned, and to increase the value that POEs add to the facility acquisition process.

The Federal Facilities Council (FFC) is a cooperative association of 21 federal agencies with interests and responsibilities for large inventories of buildings. The FFC is a continuing activity of the Board on Infrastructure and the Constructed Environment of the National Research Council (NRC), the principal operating agency of the National Academy of Sciences and the National Academy of Engineering. In 1986, the FFC requested that the NRC appoint a committee to examine the field and propose ways by which the POE process could be improved to better serve public and private sector organizations. The resulting report, *Post-Occupancy Evaluation Practices in the Building Process: Opportunities for Improvement*, proposed a broader view of POEs—from being simply the end phase of a building project to being an integral part of the entire building process. The authoring committee recommended a series of actions related to policy, procedures, and innovative technologies and techniques to achieve that broader view.

In 2000, the FFC funded a second study to look at the state of the practice of POEs and lessons-learned programs among federal agencies and in private, public, and academic organizations both here and abroad. The sponsor agencies specifically wanted to determine whether and how information gathered during POE processes could be used to help inform decisions made in the programming, budgeting, design, construction, and operation phases of facility acquisition in a useful and timely way. To complete this study, the FFC commissioned a set of papers by recognized experts in this field, conducted a survey of selected federal agencies with POE programs, and held a forum at the National Academy of Sciences on March 13, 2001, to address these issues. This report is the result of those efforts.

Within the context of a rapidly changing building industry and the introduction of new specialty fields and technologies into the building process and new design objectives for buildings that are sustainable, healthy, and productivity enhancing, and with ever-greater demands on limited resources, POE constitutes a potentially vital contribution in the effort to maintain quality assurance. Within the federal government, the downsizing of in-house facilities engineering organizations, the increased outsourcing of design and construction functions, and the loss of in-house technical expertise, all underscore the need for a strong capability to capture and disseminate lessons learned as part of a dynamic project delivery process. We hope this report will help federal agencies and other organizations to enhance those capabilities.

Lynda Stanley
Director, Federal Facilities Council

Contents

1

Overview: A Summary of Findings

INTRODUCTION

Post-occupancy evaluation (POE) is a process of systematically evaluating the performance of buildings after they have been built and occupied for some time. POE differs from other evaluations of building performance in that it focuses on the requirements of building occupants, including health, safety, security, functionality and efficiency, psychological comfort, aesthetic quality, and satisfaction. "Lessons learned" refers to programs aimed at collecting, archiving, and sharing information about successes and failures in processes, products, and other building-related areas for the purpose of improving the quality and life-cycle cost of future buildings. Ideally, the information gained through POEs is captured in lessons-learned programs and used in the planning, programming, and design processes for new facilities to build on successes and avoid repeating mistakes.

In 2000 the Federal Facilities Council, a cooperative association of 21 federal agencies with interests and responsibilities for large inventories of buildings, funded a study to look at the state of the practice of POEs and lessons-learned programs in federal agencies and in private, public, and academic organizations both in the United States and abroad. The primary purpose was to produce a report that identified successful post-occupancy evaluation programs (those that have worked well in terms of impact, longevity, and user satisfaction) and lessons-learned programs in federal agencies and the private sector. Specific objectives were to identify:

- an industry-accepted definition of POEs;
- methods and technologies used for data collection;

- the costs of POE surveys;
- the benefits of conducting POEs and capturing lessons;
- organizational barriers to conducting POEs;
- a standardized methodology that could be used within agencies to assure consistency in data gathering and allow for cooperative development of benchmarks and best practices; and
- performance measures for POE programs.

To produce this report the Federal Facilities Council commissioned a set of papers by recognized subject matter experts, conducted a survey of six federal agencies with POE programs, and held a forum at the National Academy of Sciences on March 13, 2001.

ORGANIZATION OF THIS REPORT

The next sections of Chapter 1 summarize the findings contained in the authored papers, the survey questionnaires, and the forum presentations as they relate to the study objectives. In Chapter 2, "The Evolution of Post-Occupancy Evaluation: Toward Building Performance and Universal Design Evaluation," Wolfgang Preiser reviews the historical development of POE programs, their uses, costs, and benefits; describes an integrative framework for building performance; and outlines a new paradigm for universal design evaluation. Chapter 3, "Post-Occupancy Evaluation: A Multifaceted Tool for Building Improvement," written by Jacqueline Vischer, discusses the historical basis for POE programs; identifies the discrepancy that exists between reasons for doing POEs and the difficulties of implementing them; describes successful POE programs employing the building-in-use assessment system; and makes recommendations for an unobtrusive POE process. Chapter 4, "Post-Occupancy Evaluation Programs in Six Federal Agencies," summarizes the survey questionnaire findings and describes current and emerging POE practices in those agencies. In Chapter 5, "Post-Occupancy Evaluations and Organizational Learning," Craig Zimring and Thierry Rosenheck identify the elements necessary for organizational learning; explore how 18 organizations have used POEs successfully for organizational learning; and discuss the lessons-learned role of POEs. Chapter 6, "The Role of Technology for Building Performance Assessments," authored by Audrey Kaplan, identifies technologies that have been used for POE processes; explores the possibilities of cybersurveys for improving the response rate and lowering the costs of POE data collection and analysis; and discusses Web survey design considerations, sampling techniques, publicity, data collection, and responses.

In Appendix A, "Functionality and Standards: Tools for Stating Functional Requirements and for Evaluating Facilities," Francoise Szigeti and Gerald Davis discuss how the *ASTM Standards on Whole Building Functionality and Serviceability* (ASTM, 2000) can be used to evaluate the quality of the performance delivered by a facility in relation to the original expectations. Appendix B, "A Balanced Scorecard Approach for Post-Occupancy Evaluation: Using the Tools of Business to Evaluate Facilities," written by Judith Heerwagen, outlines a performance-based approach that could provide an analytical structure to the entire process, from original concept through lessons learned. Appendixes C-F contain supporting materials. The bibliography is a compilation of references cited in the text and additional references submitted by the authors.

SUMMARY OF FINDINGS

Post-occupancy evaluation is based on the idea that better living space can be designed by asking users about their needs. POE efforts in Britain, France, Canada, and the United States in the 1960s and 1970s involved individual case studies focusing on buildings accessible to academic researchers, such as public housing and college dormitories. Information from occupants about their response to buildings was gathered through questionnaires, interviews, site visits, and observation; sometimes the information was linked to the physical assessment of a building. The lessons from these studies were intended to convey what design elements work well, what works best, and what should not be repeated in future buildings.

POE efforts in the United States and abroad were primarily focused on government and other public buildings from the 1960s to the mid-1980s. Private sector organizations in the United States became more actively involved with POE after the release of *Using Office Design to Increase Productivity* (Brill et al., 1985), which linked features of the office environment with worker productivity. As corporations downsized and reengineered their business processes to focus on core competencies, chief executive officers began to think of their buildings as ways to achieve such strategic goals as customer satisfaction, decreased time to market, increased innovation, attraction and retention of high-quality workers, and enhanced productivity of work groups. A number of organizations have since used POE as a tool for improving, innovating, or otherwise initiating strategic workspace changes.

Industry-Accepted Definition

As POEs have become broader in scope and purpose, POE has come to mean any activity that originates out of an interest in learning how a building performs once it is built (if and how well it has met expectations) and how satisfied building users are with the environment that has been created. POE has been seen as one of a number of practices aimed at understanding design criteria, predicting the effectiveness of emerging designs, reviewing completed designs, supporting building activation and facilities management, and linking user response to the performance of buildings. POE is also evolving toward more process-oriented evaluations for planning, programming, and capital asset management.

As a consequence, there is no industry-accepted definition for POE; nor is there a standardized method for conducting a POE. Even the term POE has come under scrutiny. Academics and others working in the field have proposed new terms, including environmental design evaluations, environmental audits, building-in-use assessments, building evaluation, facility assessment, post-construction evaluation, and building performance evaluations in an effort to better reflect the objectives and goals of POEs as they are practiced.

Methods and Technologies for Data Collection

Traditionally, POEs are conducted using questionnaires, interviews, site visits, and observation of building users. Over time, more specific processes, levels of surveys, and new technologies have been developed to better fit stakeholders' objectives and budgets. Shortcut methods have been devised that allow the researcher or evaluator to obtain valid and useful information in less time than previously.

Use of the Web and other technologies could substantially change the methods for conducting POEs and for analyzing the data generated. Web-based cybersurveys may become the primary survey instrument, owing to their convenience, low cost of distribution and return, ability to check for errors and receive data—including rich-text replies—in electronic format, and ease with which respondents can receive feedback. Two U.S. federal agencies have already begun moving in this direction. The Public Buildings Service of the General Services Administration is working with the Center for the Built Environment at the University of California, Berkeley, to develop a set of POE surveys that can be administered over the Web. Different surveys are directed to different key personnel to help determine if GSA is meeting a number of key management indicators. The Naval Facilities Engineering Command is modifying its database to integrate corporate management systems and to Web-enable its POE survey. The POE survey will draw information from the management system and alert individuals when the surveys should be administered.

For organizations seeking to link facility design and business goals, a POE approach could be used that combines assessment of the physical condition of the building and its systems with assessment of user comfort on such topics as indoor air quality and ventilation rates, lighting levels and contrast conditions, building (not occupant) noise levels, and indoor temperature (thermal comfort). Results from subjective or instrument measures could be plotted on floor plans using geographical information systems. The data could then be analyzed individually or as overlays showing the spatial distribution of a range of factors. For example, ratings of thermal comfort could be assessed with temperature data and spatial location. Occupants' perceptions of interior environments could also be linked with data from building control systems, local weather conditions, or facility usage as recorded by building-access *smart cards*. The Disney Corporation and the World Bank both have linked POE data to their geographic information systems for future planning and design purposes.

Costs of Post-Occupancy Evaluation Surveys

Depending on the type of survey conducted and the level of analysis used, the cost for a POE survey can range from a few thousand dollars per facility to U.S. $2.50 or more per square foot of space evaluated. Federal agencies have reported costs ranging from $1,800 for a simple standard questionnaire that could be completed in one hour to $90,000 for an in-depth analysis, including several days of interviews, the use of multidisciplinary teams, site visits, and report writing. Today the range of methods for conducting POEs allows an organization to tailor the technique to its objectives and available resources (time, staff, and money). Web-enabled surveys are emerging, and these may provide another technique that can be used at a relatively low cost.

Benefits of Conducting Post-Occupancy Evaluations and Capturing Lessons

Stakeholders in buildings include investors, owners, operators, designers, contractors, maintenance personnel, and users or occupants. A POE process that captures lessons can serve many purposes and provide many benefits, depending on a stakeholder's goals and objectives. These include the following:

- support of policy development as reflected in design and planning guides. The validity of underlying premises used in recurrent designs can be tested and evolutionary improvements to programming and design criteria can be identified and incorporated into standards and guidance literature.
- provision to the building industry of information about buildings in use by improving the measurement of building performance by quantifying occupant perceptions and physical environmental factors.
- testing of new concepts to determine how well they work in occupied buildings.
- generation of information needed to justify major expenditures and inform future decisions. Information generated by POEs can be used for decision-making in the pre-design phase of a new

project to avoid repeating past mistakes. It can also be used to educate decision makers about the performance implications of design changes dictated by budget cuts and to improve the way space is used as determined by stakeholders or documented standards.

- improvement of building performance throughout the life cycle. POEs can be used to identify and remediate such problems associated with new buildings as uncontrolled leakage, deficient air circulation, poor signage, and lack of storage. For facilities that incorporate the concept of adaptability, where changes are frequently necessary, regularly conducted POEs can contribute to an ongoing process of adapting the facility to changing organizational needs.
- making design professionals and owners accountable for building performance. POEs can be used to measure the functionality and appropriateness of a design and establish conformance with explicit and comprehensive performance requirements as stated in the functional program. They can also serve as a mechanism to monitor a building's quality and to notify decision makers when the performance of a building does not reach an agreed standard.
- aiding communications among stakeholders such as designers, clients, facility managers, and end users. Through active involvement in the evaluation process, the attitude of building occupants can be improved and proactive facility management that responds to building users' values can be facilitated.

Barriers to Conducting Post-Occupancy Evaluations

Despite these benefits, only a limited number of large organizations and institutions have active POE programs. Relatively few organizations have fully incorporated lessons from POE programs into their building delivery processes, job descriptions, or reporting arrangements. One reason for this limited use is the nature of POE itself, which identifies both successes and failures. Most organizations do not reward staff or programs for exposing shortcomings. In addition, relatively few organizations have created appropriate conditions for learning (i.e., conditions that allow the organization to constantly improve the way it operates under routine conditions and to respond to change quickly and effectively when the need arises).

Additional barriers to more effective use of POE and lessons-learned programs include the following:

- the difficulty of establishing a clear causal link between positive outcomes and the physical environment. This lack of a clear link can make it difficult for POE proponents to convince decision makers that the benefits received will justify the expenditure of time and money on the evaluations.
- reluctance by organizations and building professionals to participate in a process that may expose problems or failures or may be used as a method to focus (or deflect) blame. For federal agencies, senior executives may be concerned that identifying problems may be considered a weakness by Congress or the inspector general.
- fear of soliciting feedback from occupants on the grounds that both seeking and receiving this type of information may obligate an organization to make costly changes to its services or to the building itself.
- lack of participation by building users. In some cases the reluctance to participate can be attributed directly to uncertainty about senior management's commitment to the program, which may be manifested by lack of resources or visible endorsement of the program.
- failure to distribute information resulting from POEs to decision makers and other stakeholders.
- pressure to meet design and construction deadlines, which can create a time barrier to sustained POE activity. Staffs may be so focused on future projects and ongoing construction that POEs for completed and occupied projects are given lower priority.
- lack of in-house staff having the wide range of skills and technical expertise needed to direct and manage the results of evaluations and to communicate the information so that it is useful and nonthreatening. Organizations may be reluctant to hire consultants to conduct and analyze POEs if resources are limited and there is a lack of executive-level commitment to such programs. For federal agencies it may be difficult to obtain or earmark the funding needed to conduct POEs regardless of whether they are using consultants or in-house staff.
- organizational structures can create barriers when responsibilities for POE administration and

lessons-learned database development are assigned to different offices, thereby creating a need for interoffice collaboration and blurring the lines of accountability.

Successful Post-Occupancy Evaluation and Lessons-Learned Programs

Despite the barriers mentioned above, POE has continued to grow as a practice. Some organizations have been able to effectively integrate the lessons of POEs into strategic planning and decision-making processes for facility delivery and acquisition. Notable examples include the following:

- the U.S. Army Corps of Engineers program of participatory programming and evaluation during the 1970s: The effort, undertaken after research indicated that aging facilities were an impediment to recruiting and retaining soldiers for the all-volunteer Army, resulted in design guides for facilities ranging from drama and music centers to barracks and military police stations.
- the U.S. Postal Service program: In the 1980s the newly reorganized U.S. Postal Service made extensive use of POEs to gather information about buildings to develop retail-focused postal stores to better compete with private sector companies. The program proved successful in meeting its objectives. Over time the survey methodologies have been modified to support new objectives, but the Postal Service program remains active.
- the Disney Corporation, which uses three evaluation programs and corresponding databases to explore optimal conditions and develop predictors of its key business driver, intention of the customer to return: The databases are used extensively in the design and renovation of buildings because they allow Disney to make direct links between inputs, such as proposed number of people entering gates, and outputs, such as the width of Main Street.
- the Massachusetts Division of Capital Planning and Operations, which links POE with pre-design programming of public buildings: POEs have been used to develop and test prototype concepts for state police stations, child care centers, and vehicle maintenance and repair stations, resulting in savings of cost and time in the programming, design, and construction of new facilities.

- Bell Canada and the World Bank: Both organizations have attempted to use POE as an asset management tool for space planning. Both companies collected large amounts of data from occupants and calculated baseline scores on seven comfort factors across all buildings. Real estate staff could then identify which spaces exceeded baseline scores and which fell below. These factors are considered in budgeting for maintenance and repair, space planning, and the reconfiguration of space. The World Bank also made an effort to link its database with computer-aided design drawings so that baseline scores of buildings, or floors of buildings, could be used as indexes of quality in its space-planning process.

Elements for Post-Occupancy Evaluation Success

One objective of this study was to identify a standardized methodology that could be used by federal agencies to assure consistency in data gathering and allow for cooperative development of benchmarks and best practices. As outlined above, POEs can serve a variety of purposes and the methods used for POE surveys can be tailored to the specific purpose and available resources. It is not clear that a standardized methodology for POEs that could be used for benchmarking across federal agencies would be effective or even desirable. However, based on the information in the following chapters, it is evident that organizations seeking to establish or restructure POE programs need to make a number of key decisions in the early planning stages and develop and incorporate several key components in their program if they are to be successful, regardless of the POE purpose or methodology. These decisions and components are identified below.

- Develop a clear statement about what the organization wants to achieve by conducting and applying POEs. The links between evaluations and stated requirements have to be explicit and easy to trace.
- Identify the resources available to carry out the POE, matching data collection and analysis activities to the available time and budget.
- Identify the likely users of POE results and determine how they need the results communicated.
- Garner support from high levels of the organization to signal the importance of the project or program to people throughout the organization.

- Determine if it is to be a one-of-a-kind case study or a standardized approach that allows building professionals to collect modest amounts of comparable data from a variety of buildings on a comparative basis over time.
- For a standard approach, capture and present the information that is fed forward from such activities in comparable formats. Develop accepted terminology, standard definitions, and normalized documentation at the outset to make such comparisons easier.
- Have the questionnaire designed and analyzed by someone skilled in survey research. Determine what indicators will be selected for measurement. Evaluate such measures for usefulness, reliability, validity (they should measure what they purport to measure), efficiency, ability to allow small changes to be noticed, and balance. The entire set of measures should include both quantitative and qualitative measures, as well as direct and indirect measures.
- Consider POE techniques that avoid direct questioning of users, for example, using data generated by building control systems, observations, and expert walk-throughs. The evaluation process will fail if occupants are reluctant to participate or if staff resources are insufficient to help with the organizational data gathering or for other measures.
- Decide whether user survey data will be made available to building occupants and, if so, in how much detail and for what purpose.
- Inform facility occupants about the purpose of their involvement in providing feedback and how the data will be used. They need to be assured that their own input will be kept confidential and should be informed if immediate correction of problems is not envisioned. Occupants are more likely to be engaged in the process if they are involved in helping design the measurement plan and if they see a benefit from participation.

Elements for Successful Lessons-Learned Programs

Issues related to applying lessons learned from POEs are only partly technical. Tools for creating Web sites and databases are now widely available and relatively inexpensive. Optimizing the value of POEs requires the initiative to collect the information, the time to make sense of it, and the will to share it. To success-

fully incorporate the lessons from POEs into capital asset management processes and decision making, organizations need to do the following:

- Gain the support and long-term commitment of senior management.
- Create broad opportunities for participation and reflection. Encourage the direct participation of building owners, decision makers, designers, customers, project managers, and staff in the evaluations. Where necessary, create incentives for participation, for instance, through contract clauses, "free" vacation days, or using POE results as part of a review of qualifications when selecting consultants and contractors.
- Provide access to the information for different audiences. Upper-level management, architects, engineers, project managers, clients, and building users have different levels of responsibility in the building process and require different information from POEs. Lessons learned through POEs should be presented in a variety of formats to meet the needs of various stakeholders; these formats can include databases, design guides, case study reports, and policy and planning documents.
- Create simple databases that can be accessed by key words and that use simple methods to display overall results to aid interpretation. Ideally, a database should include the design hypotheses and assumptions for each project, the specific measures used to test the hypothesis, before and after photos of the space, brief summaries of the data, some analyses of cost, size and materials, key lessons learned, connections to other studies, connections to the full research findings before and after, and recommendations for future designs.
- Identify the critical points in the building process at which information or a POE can help resolve a problem or issue of considerable importance to participants. Use the information gathered to develop or modify policy documents.
- Build on facility evaluations that are the subject of complaints or controversy. Focus on understanding the reasons for problems or failures, and use this information to modify programming or planning processes to avoid repeating the experience.
- Use POEs of innovative buildings or components to help decide whether such innovations should be considered for future buildings.

- Create protected opportunities for innovation and evaluation. Methods for doing this include sanctioning research with the clear understanding that not all innovations will be successful, starting small with projects that have an experimental component, and evaluating the results before applying them on a broader basis.

Performance Measures

The use of performance-based approaches to facility acquisition and evaluation is a worldwide trend. Performance-based approaches require greater attention to the definition and description of purposes (demand and results) of a project or program, both in the short and long term, and to ways of measuring whether the desired results have been achieved.

Performance criteria for POEs for individual buildings are based typically on the stated design intent and criteria contained in or inferred from a functional program. Measures include indicators related to organizational and occupant performance, such as worker satisfaction and productivity, and safety and security, but may also include measures of building performance as perceived by users such as air quality, thermal comfort, spatial comfort, ergonomics, privacy, lighting comfort, noise (from the building and offices), and aesthetics.

A performance-based approach that could be used to measure the quality of services delivered by the facility in support of individuals or groups involves the use of scales created by the International Centre for Facilities. The scales have been approved and published by the American Society for Testing and Materials (ASTM) as *ASTM Standards on Whole Building Functionality and Serviceability* (ASTM, 2000). The ASTM standard scales include two matched, multiple-choice questionnaires. One questionnaire, used for setting workplace requirements for functionality and quality, describes customer needs (demand) as the core of front-end planning. The other, matching questionnaire is used for assessing the capability of a building or design to meet these levels of need, which represents its serviceability. It rates facilities (supply) in terms of performance as a first step toward an outline performance specification. In the pre-project planning phase, the scales can be used to determine if the proposed facility will meet the needs of the customer and, if not, the changes that can be made to improve the fit. Once the facility is built a second evaluation can be made to determine how closely the facility fits the original expectations.

Evaluating the performance of buildings as a financial asset within a portfolio or inventory of buildings is more difficult. One potential approach is the Balanced Scorecard system of measurement adapted to facilities. The Balanced Scorecard is a business tool that assesses four categories of performance: financial, business process, customer relations, and learning and growth (human resource development). The four categories are used to avoid overemphasis on financial incomes, to capture the full value of the product or process, and to balance levels of analysis from individual and group outcomes to higher-level organizational outcomes. Performance measures for facilities often overemphasize costs because there are few data to show linkages between facility design and business goals. Cost-focused strategies include reducing the size of work stations, using a universal plan with only a few work station options, eliminating private offices or personally assigned spaces, and telecommuting. A Balanced Scorecard approach, adapted to facilities, can answer questions such as the following:

- How can workplace design positively influence outcomes that organizations value?
- How can workplace design reduce costs or increase revenue?
- How can workplace design enhance human resource development?
- How can the physical workplace enhance work processes and reduce time to market?
- How can the physical work environment enhance customer relationships and present a more positive face to the public?

By asking these questions at the beginning of a design project the Balanced Scorecard approach can provide an analytical structure to the entire process, from conceptualization through evaluation and finally to lessons learned. POE results can be used to develop measures for all categories of performance and to evaluate the organization's success in meeting its performance goals. A core set of measures can be used across facilities to gain a better understanding of the entire building stock, while other measures would be unique to the goals and objectives of the particular organization, department, or division.

POEs and Lessons-Learned Programs for Federal Agencies

Federal agencies that own, provide, use or manage large inventories of facilities face a number of challenges. They are responsible for delivering buildings that are safe, secure, sustainable, accessible, cost effective to operate and maintain, responsive to customer needs, and supportive of worker productivity. Agencies must also become more business-like in their practices, more accountable to the public, and to manage their processes such that the results are measurable (i.e., on time, within budget, and of a quality to last 50 years or longer). In most agencies, the staff resources to meet these challenges were reduced during the 1990s and agencies face further loss of technical expertise through retirements and attrition.

In this environment, POE and lessons-learned programs, appropriately designed and managed, can add significant value to federal facility acquisition processes. A constructed facility is a culmination of policies, actions, and expenditures that call for evaluation. POE and lessons-learned programs can provide a systematic method for assessing the impact of past decisions and using these assessments in future decision making. They can partially mitigate the loss of in-house staff by creating an institutional data base that remains with the agency through changes in management and attrition in personnel. They also provide an opportunity to improve user satisfaction and reduce the cost of design development by sharing information throughout an agency and with outside contractors.

Although POE and lessons-learned programs have been instituted in relatively few federal agencies, those agencies have reported significant benefits. The General Services Administration, the Administrative Office of the U.S. Courts, and the U.S. State Department, among others, have used lessons from POE surveys to improve the design of federal buildings, to lower operating and maintenance costs, and to provide quality assurance. The Naval Facilities Engineering Command and the Army Corps of Engineers, among others, have used POE data to design buildings that better meet user needs and help to support the retention of military and civilian staff. The U.S. Postal Service has used POE data to better meet its customer needs and, in so doing, the USPS has been better able to compete with private sector mail delivery services. The U.S. Air Force and others have used POE data for quality assurance in identifying building defects and repairing them early in the life of the building when it is most cost-effective to make such repairs. The GSA and NAVFAC are restructuring their programs to better link POE to a wide range of facility management processes as federal agencies shift their emphasis from facility engineering to capital asset management.

As Web-based surveys, building controls, and geographic information systems continue to evolve, conducting POE programs and capturing lessons should become easier and yield more useful data. Instituting these programs in additional federal agencies in accord with the guidelines highlighted above should result in benefits that outweigh the costs. However, establishment of these programs will require leadership from both senior executives and program managers and a willingness to learn from both successes and failures.

REFERENCES

ASTM (American Society for Testing and Materials). 2000. *ASTM Standards on Whole Building Functionality and Serviceability*. West Conshohocken, Pa.: ASTM.

Brill, M., S.M., Margulis, and E., Konar, 1985. *Using Office Design to Increase Productivity* (2 vols.). Buffalo, N.Y.: BOSTI and Westinghouse Furniture Systems.

2

The Evolution of Post-Occupancy Evaluation: Toward Building Performance and Universal Design Evaluation

Wolfgang F.E. Preiser, Ph.D., University of Cincinnati

The purpose of this chapter is to define and provide a rationale for the existence of building performance evaluation. Its history and evolution from post-occupancy evaluation over the past 30 years is highlighted. Major methods used in performance evaluations are presented and the estimated cost and benefits described. Training, opportunities and approaches for building performance evaluation are enumerated. Possible opportunities for government involvement in building performance evaluation are sketched out. The next step and new paradigm of universal design evaluation is outlined. Last but not least, questions and issues regarding the future of building performance evaluation are raised.

POST-OCCUPANCY EVALUATION: AN OVERVIEW

A definition of post-occupancy evaluation was offered by Preiser et al. (1988): post-occupancy evaluation (POE) is the process of evaluating buildings in a systematic and rigorous manner after they have been built and occupied for some time. The history of POE was also described in that publication and was summarized by Preiser (1999), starting with one-off case study evaluations in the late 1960s and progressing to systemwide and cross-sectional evaluation efforts in the 1970s and 1980s. While these evaluations focused primarily on the performance of buildings, the latest step in the evolution of POE toward building performance evaluation (BPE) and universal design evaluation (UDE) is one that emphasizes a holistic, process-oriented approach to evaluation. This means that not only facilities, but also the forces that shape them (political, economic, social, etc.), are taken into account. An example of such process-oriented evalua-

tions was the development of the Activation Process Model and Guide for hospitals of the Veterans Administration (Preiser, 1997). In the future, one can expect more process-oriented evaluations to occur, especially in large government and private sector organizations, which operate in entire countries or globally, respectively.

Many actors participate in the use of buildings, including investors, owners, operators, maintenance staff, and perhaps most important of all, the end users (i.e., actual persons occupying the building). The focus of this chapter is on occupants and their needs as they are affected by building performance and on occupant evaluations of buildings. The term evaluation contains the world "value"; thus, occupant evaluations must state explicitly whose values are referred to in a given case. An evaluation must also state whose values are used as the context within which performance will be tested. A meaningful evaluation focuses on the values behind the goals and objectives of those who wish their buildings to be evaluated, in addition to those who carry out the evaluation.

There are differences between the quantitative and qualitative aspects of building performance and the respective performance measures. Many aspects of building performance are in fact quantifiable, such as lighting, acoustics, temperature and humidity, durability of materials, amount and distribution of space, and so on. Qualitative aspects of building performance pertain to the ambiance of a space (i.e., the appeal to the sensory modes of touching, hearing, smelling, and kinesthetic and visual perception, including color). Furthermore, the evaluation of qualitative aspects of building performance, such as aesthetic beauty or visual compatibility with a building's surroundings, is somewhat more difficult and subjective and less reliable. In other cases, the expert evaluator will pass judgment. Examples are the expert ratings of scenic and architectural beauty awarded chateaux along the Loire River in France, as listed in travel guides. The higher the apparent architectural quality and interest of a building, the more stars it will receive. Recent advances in the assessment methodology for visual aesthetic quality of scenic attractiveness are encouraging. It is hoped that someday it will be possible to treat even this elusive domain in a more objective and quantifiable manner (Nasar, 1988).

POE is not the end phase of a building project; rather, it is an integral part of the entire building delivery process. It is also part of a process in which a POE expert draws on available knowledge, techniques, and instruments in order to predict a building's likely performance over a period of time.

At the most fundamental level, the purpose of a building is to provide shelter for activities that could not be carried out as effectively, or carried out at all, in the natural environment. A building's performance is its ability to accomplish this. POE is the process of the actual evaluation of a building's performance once in use by human occupants.

A POE necessarily takes into account the owners', operators', and occupants' needs, perceptions, and expectations. From this perspective, a building's performance indicates how well it works to satisfy the client organization's goals and objectives, as well as the needs of the individuals in that organization. A POE can answer, among others, these questions:

- Does the facility support or inhibit the ability of the institution to carry out its mission?
- Are the materials selected safe (at least from a short-term perspective) and appropriate to the use of the building?
- In the case of a new facility, does the building achieve the intent of the program that guided its design?

TYPES OF EVALUATION FOR BUILDING PROJECTS

Several types of evaluation are made during the planning, programming, design, construction, and occupancy phases of a building project. They are often technical evaluations related to questions about the

materials, engineering, or construction of a facility. Examples of these evaluations include structural tests, reviews of load-bearing elements, soil testing, and mechanical systems performance checks, as well as post-construction evaluation (physical inspection) prior to building occupancy.

Technical tests usually evaluate some physical system against relevant engineering or performance criteria. Although technical tests indirectly address such criteria by providing a better and safer building, they do not evaluate it from the point of view of occupant needs and goals or performance and functionality as they relate to occupancy. The client may have a technologically superior building, but it may provide a dysfunctional environment for people.

Other types of evaluations are conducted that address issues related to operation and management of a facility. Examples are energy audits, maintenance and operation reviews, security inspections, and programs that have been developed by professional facility managers. Although they are not POEs, these evaluations are relevant to questions similar to those described above.

The process of POE differs from these and technical evaluations in several ways:

- A POE addresses questions related to the needs, activities, and goals of the people and organization using a facility, including maintenance, building operations, and design-related questions. Other tests assess the building and its operation, regardless of its occupants.
- The performance criteria established for POEs are based on the stated design intent and criteria contained in or inferred from a functional program. POE evaluation criteria may include, but are not solely based on, technical performance specifications.
- Measures used in POEs include indices related to organizational and occupant performance, such as worker satisfaction and productivity, as well as measures of building performance referred to above (e.g., acoustic and lighting levels, adequacy of space and spatial relationships).
- POEs are usually "softer" than most technical evaluations. POEs often involve assessing psychological needs, attitudes, organizational goals and changes, and human perceptions.
- POEs measure both successes and failures inherent in building performance.

PURPOSES OF POEs

A POE can serve several purposes, depending on a client organization's goals and objectives. POE can provide the necessary data for the following:

- To measure the functionality and appropriateness of design and to establish conformance with performance requirements as stated in the functional program. A facility represents policies, actions, and expenditures that call for evaluation. When POE is used to evaluate design, the evaluation must be based on explicit and comprehensive performance requirements contained in the functional program statement referred to above.
- To fine-tune a facility. Some facilities incorporate the concept of "adaptability," such as office buildings, where changes are frequently necessary. In that case, routinely recurring evaluations contribute to an ongoing process of adapting the facility to changing organizational needs.
- To adjust programs for repetitive facilities. Some organizations build what is essentially the identical facility on a recurring basis. POE identifies evolutionary improvements in programming and design criteria, and it also tests the validity of underlying premises that justify a repetitive design solution.
- To research effects of buildings on their occupants. Architects, designers, environment-behavior researchers, and facility managers can benefit from a better understanding of building-occupant interactions. This requires more rigorous scientific methods than design practitioners are normally able to use. POE research in this case involves thorough and precise measures and more sophisticated levels of data analysis, including factor analysis and cross-sectional studies for greater generalizability of findings.
- To test the application of new concepts. Innovation involves risk. Tried-and-true concepts and ideas can lead to good practice, and new ideas are necessary to make advances. POE can help determine how well a new concept works once applied.
- To justify actions and expenditures. Organizations have greater demands for accountability, and POE helps generate the information to accomplish this objective.

TYPES OF POEs

Depicted in Figure 2-1 is an evolving POE process model showing three levels of effort that can be part of a typical POE, as well as the three phases and nine steps that are involved in the process of conducting POEs:

- Indicative POEs give an indication of major strengths and weaknesses of a particular building's performance. They usually consist of selected interviews with knowledgeable informants, as well as a subsequent walk-through of the facility. The typical outcome is awareness of issues in building performance.
- Investigative POEs go into more depth. Objective evaluation criteria either are explicitly stated in the functional program of a facility or have to be compiled from guidelines, performance standards, and published literature on a given building type. The outcome is a thorough understanding of the causes and effects of issues in building performance.
- Diagnostic POEs correlate physical environmental measures with subjective occupant response measures. Case study examples of POEs at these three levels of effort can be found in Preiser et al. (1988). The outcome is usually the creation of new knowledge about aspects of building performance.

The three phases of the post-occupancy evaluation process model are (1) planning, (2) conducting, and (3) applying. The planning phase is intended to prepare the POE project, and it has three steps: (1) reconnaissance and feasibility, (2) resource planning, and (3) research planning. In this phase, the parameters for the POE project are established; the schedule, costs, and manpower needs are determined; and plans for data collection procedures, times, and amounts are laid out.

Phase 2—conducting—consists of (4) initiating the on-site data collection process, (5) monitoring and managing data collection procedures, and (6) analyzing data. This phase deals with field data collection and methods of ensuring that preestablished sampling procedures and data are actually collected in a manner that is commensurate with the POE goals.

Furthermore, data are analyzed in preparation for the final phase—applying. This phase contains steps (7) reporting findings, (8) recommending actions, and

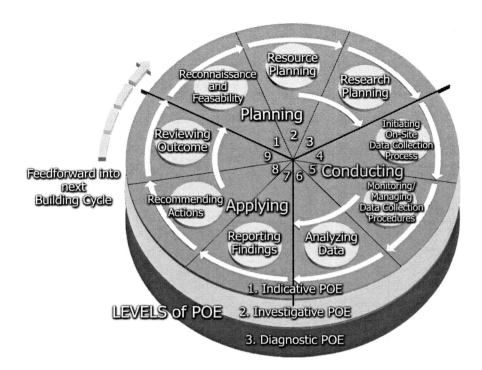

FIGURE 2-1 Post-occupancy evaluation: evolving performance criteria.

finally, (9) reviewing outcomes. Obviously, this is the most critical phase from a client perspective, because solutions to identified problems are outlined and recommendations are made for actions to be taken. Furthermore, monitoring the outcome of recommended actions is a significant step, since the benefits and value of POEs are established in this final step of the applying phase.

Critical in Figure 2-1 is the arrow that points to "feedforward" into the next building cycle. Clearly, one of the best applications of POE is its use as input into the pre-design phases of the building delivery cycle (i.e., needs analysis or strategic planning and facility programming).

BENEFITS, USES, AND COSTS OF POEs

Each of the above types of POEs can result in several benefits and uses. Recommendations can be brought back to the client, and remodeling can be done to correct problems. Lessons learned can influence design criteria for future buildings, as well as provide information to the building industry about buildings in use.

This is especially relevant to the public sector, which designs buildings for its own use on a repetitive basis.

The many uses and benefits—short, medium, and long term—that result from conducting POEs are listed below. They refer to immediate action, the three- to five-year intermediate time frame, which is necessary for the development of new construction projects, and the long-term time frame ranging from 10 to 25 years, which is necessary for strategic planning, budgeting, and master planning of facilities. These benefits provide the motivation and rationale for committing to POE as a concept and for developing POE programs.

Short-term benefits include the following:

- identification of and solutions to problems in facilities,
- proactive facility management responsive to building user values,
- improved space utilization and feedback on building performance,
- improved attitude of building occupants through active involvement in the evaluation process,
- understanding of the performance implications of

changes dictated by budget cuts, and

- better-informed design decision-making and understanding of the consequences of design.

Medium-term benefits include the following:

- built-in capacity for facility adaptation to organizational change and growth over time, including recycling of facilities into new uses,
- significant cost savings in the building process and throughout the life cycle of a building, and
- accountability for building performance by design professionals and owners.

Long-term benefits include the following:

- long-term improvements in building performance,
- improvement of design databases, standards, criteria, and guidance literature, and
- improved measurement of building performance through quantification.

The most important benefit of a POE is its positive influence upon the delivery of humane and appropriate environments for people through improvements in the programming and planning of buildings. POE is a form of product research that helps designers develop a better design in order to support changing requirements of individuals and organizations alike.

POE provides the means to monitor and maintain a good fit between facilities and organizations, and the people and activities that they support. POE can also be used as an integral part of a proactive facilities management program.

Based on the author's experience in conducting POEs at different levels of effort (indicative, investigative, and diagnostic) and involving different levels of sophistication and manpower, the estimated cost of these POEs ranges from 50 cents a square foot for indicative-type POEs to anywhere from $2.50 upward at the diagnostic level. Some diagnostic-type POEs have cost hundreds of thousands of dollars; such as those commissioned by the U.S. Postal Service (Farbstein et al., 1989). On the other hand, indicative POEs, if carried out by experienced POE consultants, can cost as little as a few thousand dollars per facility and can be concluded within a matter of a few days, involving only a few hours of walk-through activity on site.

The range of charges for investigative-type POEs,

according to the author's experience, is between $15,000 and $20,000 and covers just about that many square feet (Preiser and Stroppel, 1996; Preiser, 1998), amounting to approximately $1.00 per square foot evaluated.

The three-day POE workshop format developed by the author typically costs around $5,000, plus expenses for travel and accommodation (Preiser, 1996), and it has proven to be a valuable training and fact-finding approach for clients' staff and facility personnel.

AN INTEGRATIVE FRAMEWORK FOR BUILDING PERFORMANCE EVALUATIONS

In 1997, the POE process model was developed into an integrative framework for building performance evaluation (Preiser and Schramm, 1997), involving the six major phases of the building delivery and life cycles (i.e., planning, programming, design, construction, occupancy, and recycling of facilities). In the following material, the integrative framework for building performance evaluation is outlined. The time dimension was the major added feature, plus internal review or troubleshooting and testing cycles in each of the six phases.

The integrative framework shown in Figure 2-2 attempts to respect the complex nature of performance evaluation in the building delivery cycle, as well as the life cycle of buildings. This framework defines the building delivery cycle from an architect's perspective, showing its cyclic evolution and refinement toward a moving target of achieving better building performance overall and better quality as perceived by the building occupants.

At the center of the model is actual building performance, both measured quantitatively and experienced qualitatively. It represents the outcome of the building delivery cycle, as well as building performance during its life cycle. It also shows the six subphases referred to above: planning, programming, design, construction, occupancy, and recycling. Each of these phases has internal reviews and feedback loops. Furthermore, each phase is connected with its respective state-of-the-art knowledge contained in building type-specific databases, as well as global knowledge and the literature in general. The phases and feedback loops of the framework can be characterized as follows:

- *Phase 1—Planning*: The beginning of the building delivery cycle is the strategic plan which

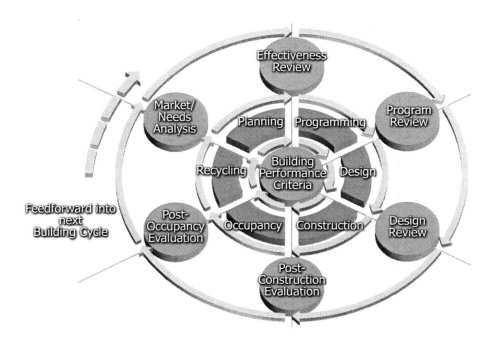

FIGURE 2-2 Building performance evaluation: integrative framework for building delivery and life cycle.

establishes medium- and long-term needs of an organization through market or needs analysis, which in turn is based on mission and goals, as well as facility audits. Audits match needed items, including space, with existing resources in order to establish actual demand.

- *Loop 1—Effectiveness Review*: Outcomes of strategic planning are reviewed in relation to big-issue categories, such as corporate symbolism and image, visibility in the context surrounding the site, innovative technology, flexibility and adaptive re-use, initial capital cost, operating and maintenance cost, and costs of replacement and recycling at the end of the useful life of a building.
- *Phase 2—Programming*: Once effectiveness review, cost estimating, and budgeting have occurred, a project has become a reality and programming can begin.
- *Loop 2—Program Review*: The outcome of this phase is marked by a comprehensive documentation of the program review involving the client, the programmer, and representatives of the actual occupant groups.
- *Phase 3—Design*: This phase contains the steps of schematic design, design development, and working drawings or construction documents.

- *Loop 3—Design Review*: The design phase has evaluative loops in the form of design review or troubleshooting involving the architect, the programmer, and representatives of the client organization. The development of knowledge-based and computer-aided design (CAD) techniques makes it possible to apply evaluations during the earliest design phases. This allows designers to consider the effects of design decisions from various perspectives, while it is not too late to make modifications in the design.
- *Phase 4—Construction*: In this phase, construction managers and architects share in construction administration and quality control to ensure contractual compliance.
- *Loop 4—Post-Construction Evaluation*: The end of the construction phase is marked by post-construction evaluation, an inspection that results in "punch lists," that is, items that need to be completed prior to commissioning and acceptance of the building by the client.
- *Phase 5—Occupancy:* During this phase, move-in and start-up of the facility occur, as well as fine-tuning by adjusting the facility and its occupants to achieve optimal functioning.
- *Loop 5—POE*: Building performance evaluation

during this phase occurs in the form of POEs carried out six to twelve months after occupancy, thereby providing feedback on what works in the facility and what does not. POEs will assist in testing hypotheses made in prototype programs and designs for new building types, for which no precedents exist. Alternatively, they can be used to identify issues and problems in the performance of occupied buildings and further suggest ways to solve these. Furthermore, POEs are ideally carried out in regular intervals, that is, in two- to five-year cycles, especially in organizations with recurring building programs.

- *Phase 6—Recycling:* On the one hand, recycling of buildings to similar or different uses has become quite common. Lofts have been converted to artist studios and apartments; railway stations have been transformed into museums of various kinds; office buildings have been turned into hotels; and factory space has been remodeled into offices or educational facilities. On the other hand, this phase might constitute the end of the useful life of a building when the building is decommissioned and removed from the site. In cases where construction and demolition waste reduction practices are in place, building materials with the potential for re-use will be sorted and recycled into new products. At this point, hazardous materials, such as chemicals and radioactive waste, are removed in order to reconstitute the site for new purposes.

UNIVERSAL DESIGN EVALUATION

The concept, framework, and evolution of universal design evaluation are based on consumer feedback-driven, preexisting, evolutionary evaluation process models developed by the author (i.e., POE and BPE). The intent of UDE is to evaluate the impact on the user of universally designed environments. Working with Mace's definition of universal design, "an approach to creating environments and products that are usable by all people to the greatest extent possible" (Mace, 1991, in Preiser, 1991), protocols are needed to evaluate the outcomes of this approach. Possible strategies for evaluation in the global context are presented, along with examples of case study evaluations that are presently being carried out. Initiatives to introduce universal design evaluation techniques in education and training programs are outlined. Exposure of students in

the design disciplines to philosophical, conceptual, methodological, and practical considerations of universal design is advocated as the new paradigm for "design of the future."

UNIVERSAL DESIGN PERFORMANCE

The goal of universal design is to achieve universal design performance of designs ranging from products and occupied buildings to transportation infrastructure and information technology that are perceived to support or impede individual, communal, or organizational goals and activities. Since this chapter was commissioned by the Federal Facilities Council, the remainder of the discussion will focus on buildings and the built environment as far as universal design is concerned.

A philosophical base and a set of objectives are the seven principles of Universal Design (Center for Universal Design, 1997).

- They define the degree of fit between individuals or groups and their environment, both natural and built.
- They refer to the attributes of products or environments that are perceived to support or impede human activity.
- They imply the objective of minimizing adverse effects of products, environments, and their users, such as discomfort, stress, distraction, inefficiency, and sickness, as well as injury and death through accidents, radiation, toxic substances, and so forth.
- They constitute not an absolute, but a relative, concept, subject to different interpretations in different cultures and economies, as well as temporal and social contexts. Thus, they may be perceived differently over time by those who interact with the same facility or building, such as occupants, management, maintenance personnel, and visitors.

The nature of basic feedback systems was discussed by von Foerster (1985): The evaluator makes comparisons between the outcomes (O) which are actually sensed or experienced, and the expressed goals (G) and expected performance criteria (C), which are usually documented in the functional program and made explicit through performance specifications. Von Foerster observed that "even the most elementary models of the signal flow in cybernetic systems require a (motor) interpretation of a (sensory) signal" and, further, "the intellectual revolution brought about by cybernetics was simply to add to a 'machine,' which was essentially a motoric power system or a sensor that can 'see' what the machine or organism is doing, and, if necessary, initiate corrections of its actions when going astray." The evolutionary feedback process in building delivery in the future is shown in Figure 2-3. The motor driving such a system is the programmer, designer, or evaluator who is charged with the responsibility of ensuring that buildings meet state-of-the-art performance criteria.

The environmental design and building delivery process is goal oriented. It can be represented by a basic system model with the ultimate goal of achieving universal design performance criteria:

1. The universal design performance framework conceptually links the overall client goals (G), namely those of achieving environmental quality, with the elements in the system that are described in the following items.
2. Performance evaluation criteria (C) are derived from the client's goals (G), standards, and state-of-the-art criteria for a building type. Universal design performance is tested or evaluated against these criteria by comparing them with the actual performance (P) (see item 5 below).
3. The evaluator (E) moves the system and refers to

such activities as planning, programming, designing, constructing, activating, occupying, and evaluating an environment or building.
4. The outcome (O) represents the objective, physically measurable characteristics of the environment or building under evaluation (e.g., its physical dimensions, lighting levels, and thermal performance).
5. The actual performance (P) refers to the performance as observed, measured, and perceived by those occupying or assessing an environment, including the subjective responses of occupants and objective measures of the environment.

Any number of subgoals (Gs) for achieving environmental quality can be related to the basic system (Preiser, 1991) through modified evaluators (Es), outcomes (Os), and performance (Ps). Thereby, the outcome becomes the subgoal (Gs) of the subsystem with respective criteria (Cs), evaluators (Es), and performance of the subsystem (Ps). The total outcome of the combined basic and subsystems is then perceived (P) and assessed (C) as in the basic system (in Figure 2-4).

PERFORMANCE LEVELS

Subgoals of building performance may be structured into three performance levels pertaining to user needs:

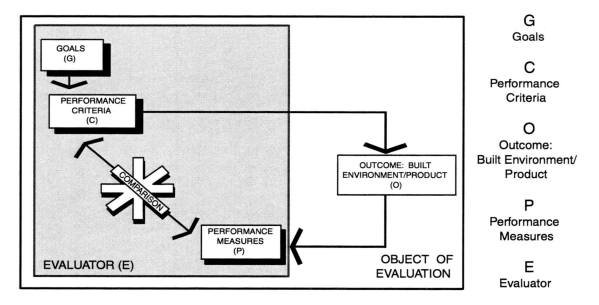

FIGURE 2-3 Performance concept/evaluation system.

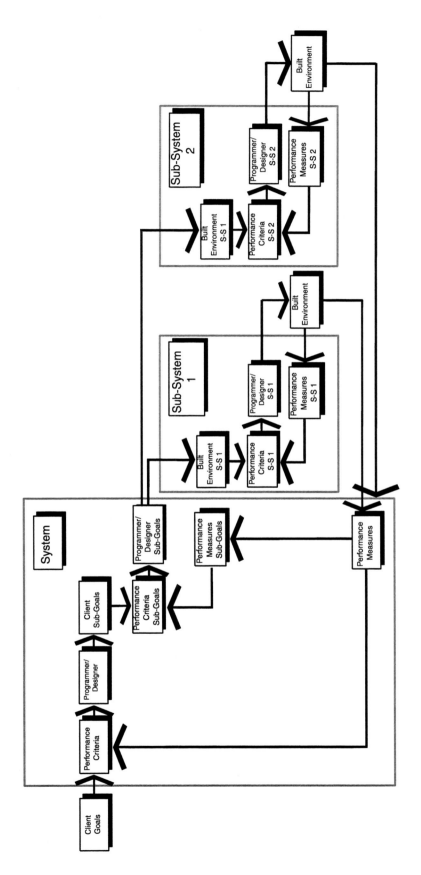

FIGURE 2-4 Feedback system with sub-systems.

the health-safety-security level, the function and efficiency level, and the psychological comfort and satisfaction level. With reference to these levels, a subgoal might include safety; adequate space and spatial relationships of functionally related areas; privacy, sensory stimulation, or aesthetic appeal. For a number of subgoals, performance levels interact and may also conflict with each other, requiring resolution.

Framework elements include products-buildings-settings, building occupants and their needs. The physical environment is dealt with on a setting-by-setting basis. Framework elements are considered in groupings from smaller to larger scales or numbers or from lower to higher levels of abstraction, respectively.

For each setting and occupant group, respective performance levels of pertinent sensory environments and quality performance criteria are required (e.g., for the acoustic, luminous, gustatory, olfactory, visual, tactile, thermal, and gravitational environments). Also relevant is the effect of radiation on the health and well-being of people, from both short- and long-term perspectives.

As indicated above, occupant needs versus the built environment or products are construed as performance levels. Grossly analogous to the human needs hierarchy (Maslow, 1948) of self-actualization, love, esteem, safety, and physiological needs, a three-level breakdown of performance levels reflects occupant needs in the physical environment. This breakdown also parallels three basic levels of performance requirements for buildings (i.e., firmness, commodity, delight), which the Roman architect Vitruvius (1960) had pronounced.

These historic constructs, which order occupant needs, were transformed and synthesized into the "habitability framework" (Preiser, 1983) by devising three levels of priority depicted in Figure 2-5:

1. health, safety, and security performance;
2. functional, efficiency, and work flow performance; and
3. psychological, social, cultural, and aesthetic performance.

These three categories parallel the levels of standards and guidance designers should or can avail themselves of. Level 1 pertains to building codes and life safety standards projects must comply with. Level 2 refers to the state-of-the-art knowledge about products, building types, and so forth, exemplified by agency-specific design guides or reference works such as *Time-Saver Standards: Architectural Design Data* (Watson et al., 1997). Level 3 pertains to research-based design guidelines, which are less codified but nevertheless of importance for building designers and occupants alike.

The relationships and correspondences between the

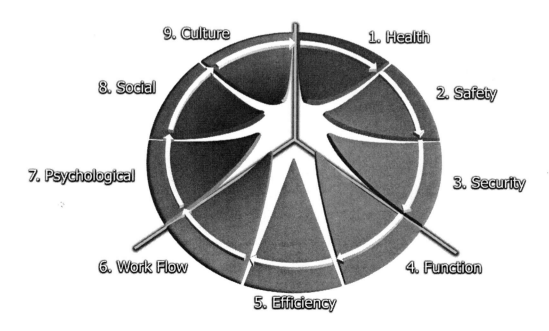

FIGURE 2-5 Evolving performance criteria.

habitability framework and the principles of universal design devised by the Center for Universal Design (1997) are shown in Figure 2-6.

In summary, the framework presented here systematically relates buildings and settings to building occupants and their respective needs vis à vis the product or the environment. It represents a conceptual, process-oriented approach that accommodates relational concepts to applications in any type of building or environment. This framework can be transformed to permit stepwise handling of information concerning person-environment relationships (e.g., in the programming specification, design, and hardware selection for acoustic privacy).

TOWARD UNIVERSAL DESIGN EVALUATION

The book *Building Evaluation Techniques* (Baird et al., 1996) showcased a variety of building evaluation techniques, many of which would lend themselves to adaptation for purposes of UDE. In that same volume, this author (Preiser, 1996) presented a chapter on a three-day POE training workshop and prototype test-

ing module, which involved both the facility planners and designers and the building occupants (after one year of occupancy), a formula that has proven to be very effective in generating useful performance feedback data. A proposed UDE process model is shown in Figure 2-7.

Major benefits and uses are well known and include, when applied to UDE, the following:

- Identify problems and develop universal design solutions.
- Learn about the impact of practice on universal design and on building occupants in general.
- Develop guidelines for enhanced universal design concepts and features in buildings, urban infrastructure, and systems.
- Create greater awareness in the public of successes and failures in universal design.

It is critical to formalize and document, in the form of qualitative criteria and quantitative guidelines and standards, the expected performance of facilities in terms of universal design.

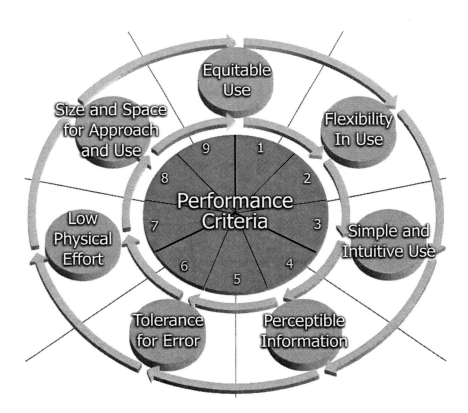

FIGURE 2-6 Universal design principles versus performance criteria.

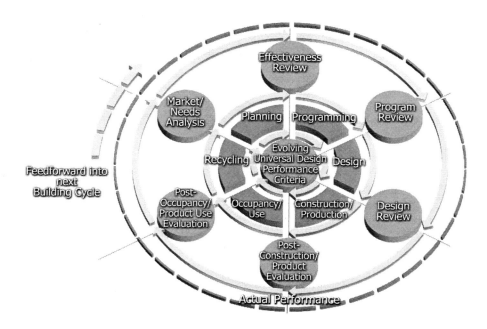

FIGURE 2-7 Universal design evaluation: process model with evolving performance criteria.

POSSIBLE STRATEGIES FOR UNIVERSAL DESIGN EVALUATION

In the above-referenced models, it is customary to include Americans with Disabilities Act (ADA) standards for accessible design as part of a routine evaluation of facilities. The ADA standards provide information on compliance with prescriptive technical standards, but say nothing about performance—how the building or setting actually works for a range of users. The principles of universal design (Center for Universal Design, 1997) constitute an idealistic, occupant need-oriented set of performance criteria and guidelines that need to be operationalized. There is also the need to identify and consider data-gathering methods that include interviews, surveys, direct observation, photography, and the in-depth case study approach, among others.

Other authors address assessment tools for universal design at the building (Corry, 2001) and urban design scales (Guimaraes, 2001; Manley, 2001). In addition, the International Building Performance Evaluation project (Preiser, 2001) and consortium created by the author has attempted to develop a universal data collection tool kit that can be applied to any context and culture, while respecting cultural differences.

The author proposes to advance the state of the art through a collection of case study examples of different building types, with a focus on universal design, including living and working environments, public places, transportation systems, recreational and tourist sites, and so forth. These case studies will be structured in a standardized way, including videotaped walk-throughs of different facility types and with various user types. The universal design critiques would focus at the three levels of performance referred to above (Preiser, 1983), i.e., (1) health, safety, and security; (2) function, efficiency, and work flow; and, (3) psychological, social, cultural, and aesthetic performance. Other POE examples are currently under development through the Rehabilitation Engineering and Research Center at the State University of New York at Buffalo. One study focuses on wheelchair users, another, on existing buildings throughout the United States. Its Web site explains that research in more detail (<www.ap.buffalo.edu/>)

Furthermore, methodologically appropriate ways of gathering data from populations with different levels of literacy and education (Preiser and Schramm, 2001) are expected to be devised. It is hypothesized that through these methodologies, culturally and contextually relevant universal design criteria will be developed over time. This argument is eloquently presented by Balaram (2001) when discussing universal design in the context of an industrializing nation such as India.

The role of the user as "user/expert" (Ostroff, 1997) should also be analyzed carefully. The process of user involvement is often cited as central to successful universal design but has not been systematically evaluated. Ringaert discusses the key involvement of the user, as noted above.

EDUCATION AND TRAINING IN UNIVERSAL DESIGN EVALUATION TECHNIQUES

Welch (1995) presented strategies for teaching universal design developed in a national pilot project involving 21 design programs throughout the United States. The initial learning from that project can be used in curricula in all schools of architecture, industrial design, interior design, landscape architecture, and urban design, when they adopt a new approach to embracing universal design as a paradigm for design in the future. In that way, students will be familiarized with the values, concept, and philosophy of universal design at an early stage, and through field exercises and case study evaluations, they will be exposed to real-life situations. As noted in Welch, it is important to have multiple learning experiences. Later on in the curriculum, these first exposures to universal design should be reinforced through in-depth treatment of the subject matter by integrating universal design into the studio courses, as well as evaluation and programming projects.

A number of authors, including Jones (2001), Pedersen (2001), and Welch and Jones (2001), offer current experiences and future directions in universal design education and training.

CONCLUSIONS

For universal design to become viable and truly integrated into the building delivery cycle of mainstream architecture and the construction industry, it will be critical to have all future students in these fields familiarized with universal design, on one hand, and to demonstrate to practicing professionals the viability of the concept through a range of POE-based UDEs, including exemplary case study examples, on the other.

The "performance concept" and "performance criteria" made explicit and scrutinized through post-occupancy evaluations have now become an accepted part of good design by moving from primarily subjective, experience-based evaluations to more objective evaluations based on explicitly stated performance requirements in buildings.

Critical in the notion of performance criteria is the focus on the quality of the built environment as perceived by its occupants. In other words, building performance is seen to be critical beyond aspects of energy conservation, life-cycle costing, and the functionality of buildings: it focuses on users' perceptions of buildings.

For data-gathering techniques for POE-based UDEs to be valid and standardized, the results need to become replicable.

Such evaluations have become more cost-effective due to the fact that shortcut methods have been devised that allow the researcher or evaluator to obtain valid and useful information in a much shorter time frame than was previously possible. Thus, the cost of staffing evaluation efforts, plus other expenses have been considerably reduced, making POEs affordable, especially at the "indicative" level described above.

ABOUT THE AUTHOR

Wolfgang Preiser is a professor of architecture at the University of Cincinnati. He has more than 30 years of experience in teaching, research, and consulting, with special emphasis on evaluation and programming of environments, health care facilities, public housing, universal design, and design research in general. Dr. Preiser has had visiting lectureships at more than 30 universities in the United States and more than 35 universities overseas. As an international building consultant, he was cofounder of Architectural Research Consultants and the Planning Research Institute, Inc., both in Albuquerque, New Mexico. He has written and edited numerous articles and books, including *Post-Occupancy Evaluation* and *Design Intervention: Toward a More Humane Architecture*. Dr. Preiser is a graduate fellow at the University of Cincinnati. He received the Progressive Architecture Award and Citation for Applied Research, and the Environmental Design Research Association (EDRA) career award. In addition, he was a Fulbright fellow and held two professional fellowships from the National Endowment of the Arts. He is a member of the editorial board of *Architectural Science Review*; associate editor of the *Journal of Environment and Behavior*; and a member and former vice-chair and secretary of EDRA. He is cofounder of the Society for Human Ecology (1978). In the mid-1980s, he chaired the National Research Council Committee on Programming Practices in the Building Process and the Committee on Post-Occupancy Evaluation Practices in the Building Process. Dr. Preiser holds a bachelor's degree in architecture from the Technical University, Vienna, Austria; a master of science in architecture from Virginia Polytechnic Institute and State University; a master of architecture from the Technical University, Karlsruhe, Germany; and, a Ph.D. in man-environment relations from Pennsylvania State University.

REFERENCES

Baird, G., et al. (Eds.) (1996). *Building Evaluation Techniques*. London: McGraw-Hill.

Balaram, S. (2001). Universal design and the majority world. In: Preiser, W.F.E., and Ostroff, E. (Eds.) *Universal Design Handbook*. New York: McGraw-Hill.

Center for Universal Design (1997). *The Principles of Universal Design* (Version 2.0). Raleigh, N.C.: North Carolina State University.

Corry, S. (2001). Post-occupancy evaluation and universal design. In: Preiser, W.F.E., and Ostroff, E. (Eds.) *Universal Design Handbook*. New York: McGraw-Hill.

Farbstein, J., et al. (1989). Post-occupancy evaluation and organizational development: The experience of the United States Post Office. In: Preiser, W.F.E. (Ed.) *Building Evaluation*. New York: Plenum.

Guimaraes, M.P. (2001). Universal design evaluation in Brazil: Developing rating scales. In: Preiser, W.F.E., and Ostroff, E. (Eds.) *Universal Design Handbook*. New York: McGraw-Hill.

Jones, L. (2001). Infusing universal design into the interior design curriculum. In: Preiser, W.F.E., and Ostroff, E. (Eds.) *Universal Design Handbook*. New York: McGraw-Hill.

Mace, R., G. Hardie, and J. Place (1991). Accessible Environments: Toward universal design. In: *Design Intervention: Toward a More Humane Architecture*. Preiser, W.F.E. Vischer, J.C. and White. E.T. (Eds.) New York: Van Nostrand Reinhold.

Manley, S. (2001). Creating an accessible public realm. In: Preiser, W.F.E., and Ostroff, E. (Eds.) *Universal Design Handbook*. New York: McGraw-Hill.

Maslow, H. (1948). A theory of motivation. *Psychological Review* 50: 370-398.

Nasar, J.L. (Ed.) (1988). *Environmental Aesthetics: Theory, Methods and Applications*. Cambridge, Mass.: MIT Press.

Ostroff, E. (1997). Mining our natural resources: the user as expert. *Innovation, The Quarterly Journal of the Industrial Designers Society of America* 16(1).

Pedersen, A. (2001). Designing cultural futures at the University of Western Australia. In: Preiser, W.F.E., and Ostroff, E. (Eds) *Universal Design Handbook*. New York: McGraw-Hill.

Preiser, W.F.E. (1983). The habitability framework: A conceptual approach toward linking human behavior and physical environment. *Design Studies* 4 (No. 2)

Preiser, W.F.E. Rabinowitz, H.Z., and White, E.T. (1988). *Post-Occupancy Evaluation*. New York: Van Nostrand Reinhold.

Preiser, W.F.E. (1991). Design intervention and the challenge of change. In: Preiser, W.F.E., Vischer, J.C., and White, E.T., (Eds.) *Design Intervention: Toward a More Humane Architecture*. New York: Van Nostrand Reinhold.

Preiser, W.F.E. (1996). POE Training Workshop and Prototype Testing at the Kaiser-Permanente Medical Office Building in Mission Viejo, California, USA. In Baird, G., et al. (Eds.) *Building Evaluation Techniques*. London: McGraw-Hill.

Preiser, W.F.E., and Stroppel, D.R. (1996). Evaluation, reprogramming and re-design of redundant space for Children's Hospital in Cincinnati. In: Proceedings of the Euro FM/IFMA Conference, Barcelona, Spain, May 5-7.

Preiser, W.F.E. (1997). Hospital activation: Towards a process model. *Facilities* 12/13; 306-315.

Preiser, W.F.E. and Schramm, U. (1997). Building performance evaluation. In: Watson, D., Crosbie, M.J. and Callendar, J.H. (Eds.) *Time-Saver Standards: Architectural Design Data*. New York: McGraw-Hill.

Preiser, W.F.E. (1998). *Health Center Post-Occupancy Evaluation: Toward Community-Wide Quality Standards*. Sao Paulo, Brazil: Proceedings of the NUTAU/USP Conference.

Preiser, W.F.E. (1999). Post-occupancy evaluation: Conceptual basis, benefits and uses. In: Stein, J.M., and Spreckelmeyer, K.F. (Eds.) *Classical Readings in Architecture*. New York: McGraw-Hill.

Preiser, W.F.E. (2001). The International Building Performance Evaluation (IBPE) Project: Prospectus. Cincinnati, OH: University of Cincinnati. Unpublished Manuscript.

Preiser, W.F.E. and Schramm, U. (2001). Intelligent office building performance evaluation in the cross-cultural context: A methodological outline. *Intelligent Building* I(1).

Vitruvius (1960). *The Ten Books on Architecture* (translated by M.H. Morgan) New York: Dover Publications.

von Foerster, H. (1985). *Epistemology and Cybernetics: Review and Preview*. Milan: Casa della Cultura.

Watson, D., Crosbie, M.J., and Callender, J.H. (Eds.) (1997). *Time-Saver Standards: Architectural Design Data*. New York: McGraw-Hill (7th Edition).

Welch, P. (Ed.) (1995). *Strategies for Teaching Universal Design*. Boston, Mass: Adaptive Environments Center.

Welch, P., and Jones, S. (2001). Teaching universal design in the U.S. In: Preiser, W.F.E., and Ostroff, E. (Eds.) *Universal Design Handbook*. New York: McGraw-Hill.

3

Post-Occupancy Evaluation: A Multifaceted Tool for Building Improvement

Jacqueline Vischer, Ph.D., University of Montreal

WHAT IS POST-OCCUPANCY EVALUATION?

Various definitions of Post-Occupancy Evaluation (POE) have been advanced over the last 20 years since the term was coined. Loosely defined, it has come to mean any and all activities that originate out of an interest in learning how a building performs once it is built, including if and how well it has met expectations and how satisfied building users are with the environment that has been created. POEs can be initiated as research (Marans and Spreckelmeyer, 1981), as case studies of specific situations (Brill et al., 1985), and to meet an institutional need for useful feedback on building and building-related activities (Farbstein and Kantrowitz, 1989). For some public agencies, such as the State of Massachusetts and Public Works Canada (now Public Works and Government Services Canada), POE is a mechanism for linking feedback on newly built buildings with pre-design decision-making; the goal is to make improvements in public building design, construction, and delivery.

Evidence from POE activities to date indicates that objectives such as finding out how buildings work once they are built—and whether the assumptions on which design, construction, and cost decisions were based are justified—are primarily of interest to large institutional owners and managers of real estate inventory. Tenant organizations, small owner-occupiers, and private sector commercial property managers are not typically investors in POE. Moreover, the number of large institutional owners and managers of real estate who have active POE programs is extremely small.

THE PROS AND CONS OF POE

One of the characteristics of POE activities is the discrepancy that exists between the reasons for doing POE (pros) and the difficulty of doing them (cons). Reasons for doing POEs are well represented in the literature. One reason is to develop knowledge about the long-term and even the short-term results of design and construction decisions—on costs, occupant satisfaction, and such building performance aspects as energy management, for example. Another reason is to accumulate knowledge so as to inform and improve the practices of building-related professionals such as designers, builders, and facility managers and even to inform the clients and users who are the consumers of services and products of those same building-related professionals. For an institutional owner-manager of real estate (government agencies, large quasi-government organizations), POE studies can provide feedback on occupant satisfaction, on building performance, and on operating costs and management practices. In sum, POE is a useful tool for improving buildings, increasing occupant comfort, and managing costs. So what mitigates against POE being a more universal activity? The barriers to widespread adoption of POE are cost, defending professional territory, time, and skills. Each one of these is examined briefly.

Cost

The cost barrier is not caused by the high costs of doing POE: building evaluation studies can be as expensive or as inexpensive as the resources available to finance them. The cost barrier is intrinsic to the structure of the real estate industry, namely, who pays for POE? In commercial real estate circles, POE is not built into the architect's fee, the construction bid, the move-in budget, or the operating budget of the building. This means that money to finance any POE activity, however small, must be found on a case-by-case basis.

Professional Territory

Defending professional territory is a barrier because POE is, after all, evaluation, and evaluation implies judgment. No active building professionals seek to have their work judged by outsiders as part of a process over which they have no control, even if the goal is a better understanding of a situation and not a performance review of a participant. It is necessary for POE to be seen as a useful *a posteriori* gathering of knowledge that is of value to the professionals involved, not as a critique of professional performance.

Time

The question of time is a mysterious one in commercial real estate. Every new building project has a rushed and constraining schedule, and every stage is carried out under unbending time pressures, although the reasons for this are not always clear at the time and in spite of the fact that rushed and fast-tracked projects often lead to costly change orders and bad long-term decisions. Going back for a follow-up look at a building, however, is not bound by the time pressures of new projects and, as a result, finds no place in the phases of a conventional building project.

Skills

Finally, what are POE skills and why is the lack of them a barrier? In spite of considerable reflection and writing by academics and researchers, there is no particular technique or tool associated with POE studies. The result is that the term itself has come to be applied to a wide range of different activities, ranging from precise cost-accounting evaluations to technical measurements of building performance to comprehensive surveys of user attitudes. Defining skills so broadly means that no one individual is likely to have all that are needed; it also means that POE does not fall into the skill set of any one individual or discipline and therefore tends to fall through the cracks.

In spite of the power of both the pros and cons of POE, the activity continues to be legitimized by one-off studies commissioned by large-scale owner-occupiers as well as by companies who, for one reason or another, are seeking to make more strategic real estate decisions. The term continues to have currency among academic researchers who are motivated to add to the general knowledge base about how buildings work after occupancy and how the environments they offer affect users. In the current context of new work patterns, changing office technology, and a more strategic approach to workspace planning, POE studies are finding a potentially valuable role in guiding companies toward more informed decision-making about office space.

CURRENT STATUS OF POE

POE has evolved from early efforts at environmental evaluation that focused on the housing needs of disadvantaged groups and efforts to improve environmental quality in government-subsidized housing. The idea that better living space could be designed by having better information from users drove environmental evaluation in Britain, France, Canada, and the United States during the 1960s and into the 1970s. It was only after the widespread acceptance of this logic—that finding out about users' needs was a legitimate aim of building research—that other building types became targets for evaluation, namely, public buildings, including courthouses, prisons, and hospitals.

The building type most recently identified as a candidate for POE is office and commercial building design. Starting with the BOSTI study (Brill et al., 1985) linking features of the office environment to employee productivity, the corporate preoccupation with reducing space costs and improving productivity has caused the private sector to become more actively involved in POE. The challenge is to build POE into the cycle of corporate real estate decision-making so that professionals involved in building programming, design, construction, and operation can acquire the relevant tools and skills; so that provision for POE is built into either the operating or the capital budget; and so that the results of a POE feed into decision-making in a useful and constructive way.

In the following sections, four types of POE are identified. Each one is illustrated with at least one case study showing how it has been used. Although these four categories of POE are not exhaustive, they seem most useful for this overview. They are

1. building-behavior research, or the accumulation of knowledge;
2. feeding into pre-design programming;
3. strategic space planning; and
4. capital asset management.

At the end, the "best practices" that can be identified from this comparative analysis are summarized. Ultimately, the *process* of POE is seen as critical in terms of meeting the challenges identified above. In the last section, a functionally viable POE process is outlined.

Building-Behavior Research, or the Accumulation of Knowledge

The notion of POE as a routine activity of the real estate industry has not gained ground in Europe, where it remains an active area of applied research in most countries. At a seminar in Paris in 1992, French policymakers, public servants, and administrators were exposed to a rich panoply of North American POE research on public buildings in order to demonstrate the value of the approach and the increasing knowledge about buildings (Centre Scientifique et technique du bâtiment, 1993). In most European circles, the idea appears to be limited to individual academic researchers who carry out housing research,[1] some office building studies,[2] and public building POEs,[3] as well as a growing number of hospital POEs in Sweden, Germany, and England (Dilani, 2000).

Funded by government agencies and using academically defensible research methods to study largely public building use, POE in Europe and Japan seems to be directed ultimately at building a broader and more reliable base of knowledge of human behavior in relation to the built environment, knowledge that may eventually come to be recognized as an academic discipline (environmental psychology, interior design?) but is not actively channeled to designers or other professionals in the real estate industry.

POE was identified as a component of the Project Delivery System used by Public Works Canada in the early 1980s, and was intended as a final stage in the programming, design, construction, and occupancy process of federal projects. A multidisciplinary approach to POE was developed and implemented for a short time in different federal office buildings in Canada (Public Works Canada, 1983). Precipitated by a concern with energy consumption in the early 1980s, studies were initiated of the performance of building systems, patterns of energy use in large buildings, and effects on occupants' perceptions of comfort. These studies led to methods to devise effective but simplified data-gathering methods to provide reliable indicators of building quality (Ventre, 1988). The glue that bound these data-gathering and analysis efforts together was an analysis of user behavior and the links that could or could not be made with building operations.

A technique for assessing user comfort was one tool that emerged out of the Canadian effort and has since been widely implemented in private industry (Dillon and Vischer, 1988). An extensive survey of users was initiated in some eight government buildings in Canadian cities, and a major data analysis effort aimed to integrate the feedback from users with data collected from instruments measuring indoor air quality, thermal comfort, lighting and acoustic conditions, and energy performance. Analysis of the questionnaire results led to the conclusion that there are seven major conditions that affect users' perceptions of their comfort in office buildings; each can be related to measures of performance of technical building systems, but not in direct or causal ways (Vischer, 1989).

The identification of what came to be known as the Building-In-Use measure of ambient environmental comfort led to the development of a standardized measurement tool in the form of a short questionnaire. The questions are formatted as 5-point scales on which building occupants rate the seven key dimensions of environmental comfort in their workplace. Both the five-year data-gathering and analysis effort that led up to the building-in-use system and its subsequent extensive use in the private sector can be considered a major POE initiative that has important implications for building-behavior research and has also generated a tool that can be used for other types of POE (Vischer, 1996).

Since its development for the Canadian government, the Building-In-Use (BIU) assessment system has been used all over the world. Two books in English and one in French, along with several articles, have been published that describe the system and its applications. A copy of the questionnaire is contained in Appendix C. Subsequent sections of this chapter deal with applica-

---

[1]For example, Mirella Bonnes at the University of Rome in Italy.

[2]For example, Peter Jockusch at the University of Kassal, Germany.

[3]For example, the Building Research Institute in Olso, Norway.

tions of this POE system in a variety of different contexts.

This year, a new research initiative in Quebec, Canada, has identified among other objectives the need to update and modernize the BIU assessment system. This objective is combined with another broad-ranging research goal, that of carrying out a POE of some 3,000 universal work-station installations in the offices of Quebec's largest insurance company. Not only is this POE targeting ambient environmental conditions in the work environment, as was done in the 1980s, but it will also examine the psychological impact on individual and group work of a highly standardized work environment.

This study aims to make a valuable contribution to the office POE literature by adding to existing knowledge of building systems performance and human comfort. As well as updating our knowledge of key environmental conditions in the workplace, this research will measure psychological needs such as privacy and territoriality and the influences of group norms and membership as well as organizational values on occupant perception of the work environment. More purely social science methods are being used, such as individual interviews, focus groups, and a questionnaire survey to be carried out through individual interviews with a stratified random sample of the populations of up to six different buildings. Results will become available in the fall of 2001, and the study will be published in 2002.

Linking POE with Pre-Design Programming

One of the most appealing reasons to perform POE is to be able to inform building decision-making in the early stages of a new project. POE studies target user evaluation of an existing space where users are destined to occupy a new space that is being planned. Their feedback is needed to ensure that the new design meets users' needs and solves problems in existing buildings. Certain public agencies such as the Division of Capital Planning and Operations in Massachusetts have POE as a legitimate and funded stage in all capital projects; the activity is run by the Office of Programming, the office responsible for all pre-design planning. The concept behind the legislation was to link POE with pre-design programming of public buildings.

In certain projects, the link has been effected and has paid off. For example, State Police stations are all designed along the same principles because they all serve the same functions. The Office of Programming

developed a prototype concept that was built and occupied. The post-occupancy evaluation was part of the design process for new police stations, and the prototype was carefully examined in use, with its functionality, costs, structure, and materials evaluated. The prototype design was then modified, and the design of state police stations was standardized and built along the same lines. This saved time and effort on programming, design, and construction costs for the state. A similar approach has worked for child care centers and state vehicle repair and maintenance centers, and was being considered for state armories and firing ranges.

Soon after a large sum was approved by the Massachusetts legislature for a fast-tracked program of new prison construction, a post-occupancy prison study was carried out. The results of the new Old Colony state penitentiary POE were delivered to the Department of Corrections and ultimately used in programming four to six fast-tracked corrections projects by the Office of Programming.

However, in other projects, POE was not as successful. It was not uncommon for the budgeted amount for POE to be used up by change orders and other requirements of the construction process. These projects had no funds left for the POE stage of the process. In other cases, the time barrier alluded to above created a misalignment between POE studies and programming and design activities. For example, POEs of the correctional institutions built early in the fast-tracked process were not done because their results would not have been available in time to inform programming and design for the next project. A POE on a Massachusetts courthouse in 1985-1986 was only approximately aligned with the state's courthouse construction program that was funded from 1984-1989.

POEs are still part of the public building programming process, and efforts have recently been made to develop and implement a standardized POE procedure that will fit in with the state's building programs, provide the right information at the right time, and enable project managers to identify more exactly for each project the amount of money needed for POE.

POE was also built into government building delivery processes in New Zealand in the 1980s before the privatization of the public works department. Performing POEs facilitated pre-design programming and gave the design and construction team on each project a closer contact and understanding of users in each of its projects. This led to a design approach characterized by the designers as a negotiation, with multiple exchanges of information and openness to change on both sides (Joiner and Ellis, 1989). However, this approach was threatened by the severe government cutbacks of the 1980s, closely followed by extensive privatization that put the public works departments in competition with private firms for public building projects. Although the close links with users that the public works department had developed gave it an edge, the time needed for negotiated design was not competitive, even though it could demonstrate that savings would be realized later on in the life of the building.

These accounts suggest that in spite of the logical imperative to link POE results to the front end of the design process, efforts to do so have had to struggle to survive. This should not be taken to mean that such efforts are futile; on the contrary, they are valiant and should be continued, if only to make us question the basic irrationality of the present building creation process.

POE in Strategic Space Planning

There is a clear difference between attempts to feed POE study results into pre-design decision-making on a routine basis, such as those described above, and using POE in strategic space planning. This latter use of POE has gained credibility in recent years as corporations are trying increasingly to provide functionally supportive workspace to their employees and simultaneously to reduce occupancy costs.

A number of companies in recent years have used Building-In-Use assessment to initiate a process of strategic space planning. This approach indicates a more complex situation than that which is characterized by conventional office space planning. It implies that the organization seeks to improve, innovate, or otherwise initiate workspace change to bring space use more in line with strategic business goals. These changes are not always understood or accepted by employees; thus, some companies have taken a change-management approach to new space design. Others have imposed major workspace changes in the same way as any other redesign of the workspace, sometimes with markedly negative results (Business Week, 1996).

One example of a company that used POE for strategic space planning in the early 1990s and has realized significant gains in corporate recruitment and retention, as well as increased sales, is Hypertherm Inc. (Zeisel, in press). This medium-sized manufacturing company

located in New Hampshire is a world leader in the design and production of plasma metal-cutting equipment. As a result of a need for expansion of both its manufacturing plant and its offices, Hypertherm hired an architect whose drawings and conceptual approach it later rejected on the grounds that a new work environment should accompany the significant organizational change that was needed to prepare the company for global expansion and a better competitive position as its share of the world market grew.

The workspace design process at Hypertherm included a POE of the existing space. As well as providing useful data to help design the new space, the survey caused employees to feel both consulted and involved in the process of designing new space that would meet business goals. The strategic space planning comprised a number of different steps. First, a shared vision of the new work environment was created through *a structured team walk-through* of the existing facility. The management team and consultants toured the facility as a group, discussing the tasks of each work group, pointing out difficulties and advantages with the present space, and commenting on each other's presentations. Recorded on cassette tape and transcribed, the tour commentary was presented back to the client, along with a comprehensive set of photographs, to document existing conditions. This activity involved all members of the team and gave the consultant a large amount of information efficiently; consensus began to build on what needed to be done to solve the company's space problems.

A subsequent series of *facilitated work sessions* enabled the management team to generate a set of goals and objectives not only for the new space but also for the restructured organization. This stage yielded a set of design guidelines to be applied to the new design and established priorities as to the relative importance of what the team wanted to achieve. Most importantly for the future of the process, it resulted in consensus.

The next step required *involving employees in the design process* in order to ensure widespread acceptance of the vision. The BIU assessment survey of employee perceptions of the physical conditions of the existing building started the process of employee involvement. Each employee filled out the survey and was therefore alerted to the imminent new space design process and to the importance of his or her role in it. The survey results were published in the company newsletter and showed the best and the worst aspects of working in the old space.

A final step in the strategic approach was to invite employee representatives to provide feedback on design development and to communicate key design decisions to their colleagues.

The questionnaire survey was distributed a second time, about six months after move-in, to compare levels of user comfort between the old and the new buildings. Not only were occupants pleased with their new space, but the process of buy-in and participation also helped them understand from day one how it would work. They accepted it without the discomfort and resistance often exhibited in new work environments, because they knew exactly what to expect. The employees took ownership of their new space because they had been involved in decision-making throughout.

Figure 3-1 shows employee ratings on the seven dimensions of workspace comfort in the old building and after the renovation. Having a short, standardized questionnaire to compare employee perceptions before and after a workspace change is a significant tool for POE studies. BIU assessment has been used successfully in strategic space planning for a range of companies, including Bell Sygma, Reuters, Boston Financial Group, and GTE Government Systems.

Capital Asset Management

Using POE as a tool for managing building assets is not new, but it has not been widely implemented. This is perhaps due to the diverging focus of the two activities. Asset management tends to rely on data on building operating costs, maintenance and repair needs, real estate value and market conditions, and tenant improvements. The standard POE approach provides data on user perceptions and attitudes through group feedback sessions, such as focus groups, survey questionnaires, and *in-situ* observations of user behavior. As a result, feedback from employees in corporations is often considered to be of interest more to human resources departments than to the real estate team.

However, an approach to POE that combines assessment of the physical condition of the building and building systems with user comfort assessment on such topics as indoor air quality and ventilation rates, lighting levels and contrast conditions, building (not occupant) noise levels, and indoor temperature (thermal comfort) could constitute another tool to add to those used in conventional asset management. One weakness in making this potentially fruitful link is the finding that much of the literature that has been published on specific building studies has been unable to demonstrate systematic links between the feedback users provide through questioning and data derived from instrument measures of interior building conditions (Vischer, 1993). Moreover, portfolio managers tend to steer away from POE approaches because there are few tools available (the BIU assessment questionnaire being one of the few simple ways of collecting user

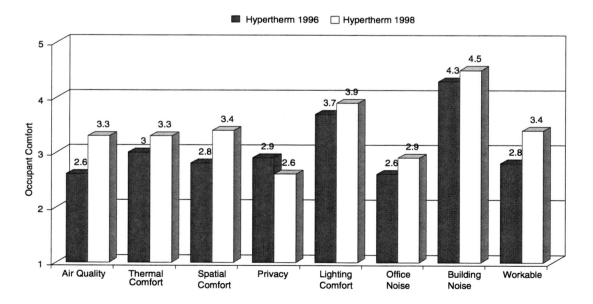

FIGURE 3-1 Before and after BIU profile comparison.

feedback on a standardized basis) and because any questioning of users requires informing and involving tenants. A more finely tuned and precise approach to POE is necessary in order to make this approach valuable to professional evaluators and asset managers in organizations.

However, BIU assessment has been used by large owner-occupier corporations to collect data on user perceptions and link these directly with ambient interior conditions and, therefore, building performance and building quality. Two examples of companies with extensive real estate holdings who attempted to use POE systematically as an asset management tool are Bell Canada and the World Bank. Both organizations owned some and leased some of their office space; both were committed to providing high-quality office space, in the first case, as part of a continuous improvement philosophy and in the second case, as part of the corporate mission; and both chose BIU assessment as their POE tool for managing assets.

In both cases, BIU surveys were carried out in almost all of their buildings, in both leased and owner-occupied space. In the case of Bell, this meant surveys of 2,500 people distributed in its headquarters tower in Montreal, in suburban office buildings in Montreal and Toronto, and in one building in Quebec City. In the case of the World Bank, this meant surveying 2,800 employees distributed in the eight leased and owned buildings it occupied in Washington, D.C., and also included one building in Paris.

Both corporations collected large amounts of feedback from occupants using the BIU questionnaire. In addition to individual building analysis, the BIU data on environmental comfort were grouped and analyzed for overall trends in occupant comfort. The large number of cases and variety of building settings surveyed enabled baseline scores to be calculated on the seven comfort dimensions across all buildings. This, in turn, allowed real estate staff to identify which buildings or parts of buildings exceeded the baseline scores and which fell below them. Both organizations found that this approach was cost-effective and not data heavy, did not consume inappropriate amounts of staff time, and provided a single-digit indicator of environmental quality. BIU results from individual buildings could easily be compared either to their own baseline, that is, the standards set by their own building stock, or to the baseline scores generated by the pre-existing BIU database, indicating a generalized North American standard of quality based on survey results from some 60 buildings.

Procedures were set in place to allow the baseline scores to be updated as new information was added through additional building surveys. One organization, Intelsat, initiated an electronic form of the questionnaire survey in order to be able to update occupant comfort ratings easily. In the case of the World Bank, an effort was made to link the BIU database to computerized drawings that were used to plan and update office layouts, so that BIU scores of buildings, floors, or areas of floors that were slated for reconfiguration could be consulted and indices of quality made available as part of the space planning process—a sort of instant POE.

BEST PRACTICES

The rapid overview of case studies presented above offers the following conclusions. First, it is clear that POEs of built environments must continue in order to enhance our knowledge about the effects of physical space on people. The challenge of the "building-behavior research" definition of POE is to ensure that the knowledge gained from research studies is not only disseminated in the academic community but also successfully transferred to the world of designers, builders, and financiers of real estate.

For public agencies or other organizations that repeatedly construct the same building type, linking POE with pre-design programming can save money and time. Evidence indicates, however, that even when the link to pre-design decision-making is recognized, POE is not simple to implement. It is likely to be successful only if, as Friedman et al. (1979) pointed out in their seminal first book on building evaluation, a structure-process approach is used. This means designing an approach ahead of time, developing and testing the process beforehand, and ensuring that resources continue to be available.

POE is also a potentially useful tool for asset management, as long as the approach employed to collect feedback from users can be effectively integrated with the other more market-oriented data-gathering efforts of asset managers. This may mean simplifying the elaborate social science approach favored by researchers and investing in a test initiative to implement and test an asset management approach to POE in the context of the real estate industry so as to demonstrate how feedback from users can both be collected easily and enhance real estate decision-making.

POE also seems to have a natural place in strategic space planning and could be developed for use by a wide variety of organizations. The key to this application is to consider POE a tool for involving building users in planning new workspace. Some organizations have reservations about allowing their employees to become too involved in the emotional and time-consuming planning of their new space. However, techniques exist for managed participation that have been successful in a variety of instances in helping to control the amount and type of user involvement. Involving users in new workspace planning is necessary for any successful change initiative, and POE is one of the tools available to this end.

Finally, in spite of the ground-breaking efforts by some large organizations to build POE techniques into their building management activities, it is curious that large property management firms are rarely known to use POEs for building diagnostic purposes or for improving services to tenants. Some property management companies content themselves with short satisfaction questionnaires that tenants complete following a repair or move. However, in the experience of this writer, property management firms, along with large banks and financial institutions, are the companies least likely to perform POE. Techniques of POE need to be developed for use by organizations that would make good use of occupant feedback if they had a simple, reliable way of getting it. These techniques would help them to build environmental evaluation into their planning, budgeting, and maintenance cycles.

It should be noted that some companies fear soliciting feedback from building occupants on the grounds that both seeking and receiving this type of information may obligate them as building owners and/or managers to make a costly change to their services or to the building itself. At least one lawsuit has been heard of, resulting from a perceived lack of follow-up to an occupant survey that questioned users about their perceptions of indoor air quality and lighting (Boston Globe, 1987).

MANAGING POE INFORMATION

Once POE exists outside the protected framework of a case study research project, another set of barriers present themselves in the form of dissemination of the information yielded by the study. As long as the POE is carried out as academic research, the sanctioned forms of academic research dissemination are available (publication in journals, conferences, etc.). However, in the practical world of building design, construction, and

management, most organizations have no established system for knowing how to process, direct, and act on the information they receive from a POE. This may cause the information not to go anywhere, and it becomes a reminder to decision-makers not to repeat the experience. Having no clear use for the information may generate conflict and resentment among those who are expected to act on it; seeing their feedback ignored and not put to good use may alienate building occupants.

Many organizations that initiate POE are unclear as to why they want the information, what information they want, to whom it should go, and how they are expected to follow up on it. Several organizations familiar to the author have explicitly required that the results—whether positive or not—of a POE survey not be disseminated.

Among the range of possible reasons for a lack of planning for the dissemination of POE information are the following:

- "usefulness" of user surveys,
- complexity of the design process,
- negativity of the comments received, and
- complexity of managing information.

Each of these is discussed below.

Usefulness of User Surveys

Questioning people about how much they like or dislike as space that they occupy inevitably obliges the researcher to confront the "so-what?" question: So what if some users like a feature and others do not? The notion of liking something is so subjective and constrained by circumstances that it is difficult to extract generic information or to generalize from users' responses. However the notion of user satisfaction is at the base of almost all POE approaches, leading to highly specific results from POE case studies and a lack of generalizable conclusions to guide additional research or changes in design (Vischer, 1985).

Some POE approaches—for example, BIU assessment—have replaced the emphasis on user satisfaction with questions that target a more functional evaluation of the work environment. BIU survey questions, for example, ask respondents to identify their level of comfort in relation to the specific tasks of their job. The intent is to shift the user feedback away from personal likes and dislikes toward what might be called an "objective" assessment of the functionality of the work environment. For example, lighting quality is rated according to the respondent's task requirements: work at a computer screen, reading print-out or other documents, or appraising forms, colors, and other visually oriented tasks. This approach is based on the concept of "functional comfort"; theoretically, any building user can evaluate functional comfort for any other person performing the same tasks in the same environment.

Organizations that have asked, "Do you like or dislike . . . ?" or "Are you satisfied or dissatisfied with . . . ?" have found the results too subjective to be useful. Managers fear that these questions raise users' expectations and cause them to expect wide-ranging corrective measures and/or gratification of their wishes. Although some POE studies target correcting problems in the building, many are initiated without a budget or a procedure for follow-up. Moreover, companies often design questionnaires with little regard to the considerations necessary for good survey design. Often such questionnaires are overly long and detailed, and analysis of the results is almost always limited to simple frequencies and percentage calculations. As a result, companies find the data less than useful and conclude that occupant surveys are a waste of time.

Complexity of the Design Process

People who are trained in and perform building design have difficulty moving beyond design into a POE once their buildings are occupied and built. Some designers evince a general curiosity that generates some unfocused evaluation activity that gives them a sense of the success of their design decisions. The feedback that the designer acquires rarely goes further than the individual seeking the information. Some designers will spend considerable time on acquiring feedback, whereas others have almost no curiosity, and most are somewhere in between. Why are design professionals not more curious to learn from the positive and negative impacts of their design decisions? I believe the answer to the question is the complexity of the design process, for the following reasons.

The approach of nondesigners to POE is somewhat different from that of designers: the former approach space as another cultural artifact to be studied using traditional social science methods. Some studies seek to orient the space evaluation to the basic design decisions and/or criteria. In the words of *Inquiry by Design* (Zeisel, 1975), each design decision is a hypothesis for

research. Other researchers have taken architects' decisions as hypotheses and tested them with POE data (Cooper, 1973). This approach implies a linear logic according to which programming (pre-design information gathering) leads to design decisions, which lead to construction of what has been designed, which in turn leads to POE. Montgomery elaborates on this linear logic in his introduction to *Architects' People* (Ellis and Cuff, 1989). As he points out, this may be a model for a rational world, but it is all too clear that the world of architecture and real estate is anything but rational, that design itself is not rational, and that trying systematically to link POE with design for those involved in that process is all but impossible.

What researchers are less aware of, and the designer is painfully aware of, is the irrational nature of design decisions. Each design decision on a project is influenced by the personality traits, role and status, and personal opinions of the individuals involved (client, project manager, architect, contractor, etc.) as well as by the stage reached at the time of making the decision, how involved and informed users are, expectations, budget, and other pressures such as government regulations, site constraints, and so forth. The designer's own design ideas, and how these are communicated, when and to whom, also affect the process. These are but a few of the factors that mean that building design is not controlled by any one person or agency and that therefore a clear notion of a project's design ideas and intentions for the purpose of POE is difficult to identify.

The author's own experience of highly participatory design processes has provided first-hand experience of the convoluted, political, and anything-but-linear decision-making process that causes a building to be what it ultimately is. In many cases, no amount of rational planning and programming can change the likelihood that once occupied, the use of the space begins immediately to change. Sometimes small adjustments are made creating incremental change over time, and sometimes the basic assumptions that guided the design of the building are dramatically forgotten and the space is adapted to serve a different purpose.

Given the nature of building design and construction processes, it is unrealistic to expect a designer to seek out feedback on the long-term effectiveness of design decisions on a systematic basis. However, this is more of a comment on the POE process than on the POE product, a product whose usefulness cannot be denied in spite of the complexity of the decisions that go into the creation of new physical environments.

Primarily Negative Feedback

Because of the emphasis on building user feedback, much of the information received from POEs is critical in nature and appears to assign inordinate weight to nonfunctional or dysfunctional aspects of a building with little mention of what works. This impression, although false, makes it difficult for information to be shared in a constructive and useful way. The best way around this dilemma is for a skilled POE researcher to weigh the importance of the information received.

For example, 10 comments from building users about dripping soap dishes in bathrooms, slippery front steps, and dust on the work surface cannot be compared in importance to one or two comments about poor lighting or a lack of meeting rooms. The former are irritants; they should not carry the weight of an item that creates a serious dysfunctionality or impedes user effectiveness. Similarly, it has frequently been this author's experience during focus group sessions with users to listen to 45 minutes of complaints and negative criticism about their space only to have them comment (once they have got all this off their chests) that they like the daylight and view from the windows, the space is much better than where they used to work, and they like working in the building!

One of the challenges of POEs going forward is to identify a reasonable system of informed weighting of user feedback so that the data received can be interpreted according to balanced positive and negative categories.

Complexity of Managing POE Information

The information that results from POE is directed in a number of different directions. In some cases, solutions are sought to problems that have been identified in a building, and the information is directed to facility managers, building owners, and landlords. In others, the information is directed to designers to help them make better design decisions on a specific project or generically with regard to a building type. In some cases, building users are informed regarding the results of a POE in which they were involved, as a way of involving them in planning change and finding ways of improving the environment. In yet other cases, the information is seen as valuable in itself and disseminated to researchers seeking to understand more about the person-environment relationship. Finally, information about building systems performance, occupant

functional comfort, operating costs, and adaptation and re-use is directed to stakeholders in the planning, design, construction, and occupancy process who are in a position to make decisions about future building projects.

For each of these, and no doubt other, applications of POE information, some thought needs to be given at the outset to collecting and presenting POE information in a way that suits the receiver and consumer of that information. This means that a clear understanding of the context is necessary and that the POE process should be designed as a function of contextual constraints. Key questions to be asked before any POE study include the following: Who wants the POE? How do they want to use the information? What resources are available to gather, analyze, and disseminate the information? Who will receive the results, and when? What expectations do stakeholders have of the POE results?

THE FUTURE OF POE: RECOMMENDATIONS FOR AN UNOBTRUSIVE POE PROCESS

The importance of the *process* used in carrying out a POE cannot be underestimated; in this author's opinion it is more important than the method selected and the data gathered. Once users are involved—as they are once they are questioned about their use and occupancy of a building—how they are approached, what information they are given, and the follow-up they experience are all critical stages in the development of the relationship between the occupant and his physical environment. The POE, then, provides an opportunity for improvement not only of the building and of the environment it provides to its users, but also of the way users perceive and feel about their territory.

Ideally, one would like to see POE carried out systematically on a wide variety of types of building, but not before clear objectives and results are identified. At the outset, it is important to clarify the value that the POE will have for the person or agency carrying it out. If one can identify that the POE has value in the context in which it is being implemented, then decisions about financing it, identifying the right things to be studied, and disseminating the results and information to the right people will follow. As the examples in this chapter show, stakeholders in private sector real estate development fail to attach value to POE, and even public agencies—for which the value of POE is apparent—

find that the complexities of the process outweigh potential gains from POE.

In conclusion, it is proposed that a workable POE process designed to succeed outside academic circles incorporate the following steps:

1. A simple, reliable and standardized way should be developed of collecting useful feedback from occupants, not on the entirety of their experience of using the building, but on a few, carefully selected and identified indicators of environmental quality.

2. The indicators selected for measurement should be decided beforehand according to that which is most relevant to the initiators of the POE and the context in which it is implemented.

3. It is necessary to clarify at the outset who are likely to be the consumers of POE results and therefore how best to communicate these results to them.

4. Consideration should be given to POE techniques that avoid direct questioning of users—for example, using instruments, observations, expert walk-throughs, etc.—as well as to refining social science techniques to devise reliable and rapid ways of questioning building occupants.

5. Efforts to combine instrument data collection and surveys of building users can be costly, because large amounts of data are generated without yielding much additional useful information. One approach is to use the analysis of user responses to indicate where and when follow-up instrument measures might clarify the nature of the problems identified and indicate possible solutions.

6. Users should be well informed regarding the purpose of their involvement is providing feedback and should be made aware in situations where immediate correction of problems is not envisioned. In fact, it is necessary to recognize that building users can be "measuring instruments" of environmental quality, rather than only customers to be served.

7. A decision should be taken at the outset as to whether or not user survey results will be made available to building occupants and, if so, in how much detail and for what purpose.

8. Resources for carrying out the POE should be defined clearly so that data collection and analysis activities fit into time and budget constraints, however modest.

9. If a questionnaire is given to occupants, it should be designed and analyzed by someone knowledgeable in survey research, even if this person is not involved in the eventual use and application of the results.

10. A standardized approach that allows building professionals (designers, developers, managers) to collect modest amounts of comparable data from a variety of buildings to analyze on a comparative basis is likely to be more useful than a detailed one-off case study in most situations.

11. Public agencies should examine the possibility of setting up test POEs on a demonstration basis, to develop POE techniques, to demonstrate value, and to determine the best ways of making POE relevant to the building industry.

ABOUT THE AUTHOR

Jacqueline Vischer is an environmental psychologist. She is currently professor and director of the interior design program at the University of Montreal. She has consulting experience in architecture and planning projects in the United States and Canada. As principal in her own consulting firm in Vancouver, Canada, Dr. Vischer and her staff undertook contract research from government agencies in residential planning and evaluation, institutional programming, and policy analysis. Dr. Vischer then spent five years developing building performance studies of office buildings for Public Works Canada in Ottawa, projects from which the building-in-use assessment system emerged. She then undertook the design and implementation of a post-occupancy evaluation program for public buildings owned and operated by the State of Massachusetts' Division of Capital Planning. In 1989 Dr. Vischer started the Institute For Building Science, which became Buildings-In-Use in 1990 and opened its Montreal office, Bâtiments-en-Usage, in 1991. Dr. Vischer has held positions as lecturer and instructor at the McGill University School of Urban Planning, University of British Columbia, and Harvard University's School of Design. She is a member of the Environmental Design Research Association, the American Society for Heating Refrigeration and Air Conditioning Engineers, the Montréal Metropolitan Energy Forum, and the International Facilities Management Association. Dr. Vischer holds a bachelor of arts in psychology from the University of California, Berkeley, and a master of arts in psychology from the

University of Wales Institute of Science and Technology, and a Ph.D. in architecture from the University of California, Berkeley.

REFERENCES

Boston Globe. (1987). 30 June.

Brill, M., Margulis, S.M., and Konar, E. (1985). *Using Office Design to Increase Productivity* (2 vols.). Buffalo, N.Y. : BOSTI and Westinghouse Furniture Systems.

Business Week. The new workplace. (1996). April 29, pp.107-117.

Centre scientifique et technique du bâtiment (1990). *Améliorer l'architecture et la vie quotidienne dans les bâtiments publics.* Paris: Plan construction et architecture, Ministère des équipements, du logement, des transports et de l'espace.

Cooper, C. (1973). *Comparison Between Architects' Intentions and Residents' Reactions, Saint Francis Place San Francisco.* Berkeley, Calif: Center for Environmental Structure.

Dilani, A. (ed.) (2000). *Proceedings of the 3rd International Conference on Health and Design.* Stockholm: University of Stockholm.

Dillon, R. and Vischer, J. (1988). *The Building in-Use Assessment Methodology* (2 volumes). Ottawa: Public Works Canada.

Ellis, W.R, and Cuff, D. (1989). *Architects' People.* New York: Oxford University Press.

Farbstein, J., and Kantrowitz, M. (1989). Post-occupancy evaluation and organizational development: the experience of the United States Postal Service. In: *Building Evaluation.* Preiser, W. (ed.). New York: Plenum Press, p. 327.

Friedman, A., Zimring, C., and Zube, E. (1979). *Environmental Design Evaluation.* New York: Plenum Press.

Joiner, D., and Ellis, P. (1989). Making POE work in an organization. In: Preiser, W. (ed.) *Building Evaluation.* New York: Plenum Press, p. 299.

Marans, R., and Spreckelmeyer, K. (1981). *Evaluating Built Environments: A Behavioral Approach.* University of Michigan, Survey Research Center and Architectural Research Laboratory.

Public Works Canada (1983). *Stage One in the Development of Total Building Performance* (12 volumes). Ottawa: Public Works Canada, Architectural and Building Sciences.

Ventre, F. (1988). Sampling building performance. Paper presented at *Facilities 2000 Symposium,* Grand Rapids, Mich.

Vischer, J. (1985). The adaptation and control model of user needs in housing. *Journal of Environmental Psychology* 5:287-298.

Vischer, J. (1989). *Environmental Quality in Offices.* New York: Van Nostrand Reinhold.

Vischer, J. (1993). Using occupancy feedback to monitor indoor air quality. *ASHRAE Transactions* 99 (Pt.2).

Vischer, J. (1996). *Workspace Strategies: Environment as a Tool for Work.* New York: Chapman and Hall.

Zeisel, J. (1975). *Inquiry by Design.* New York: Brooks-Cole.

Zeisel, J. (in press). *Inquiry by Design,* 2nd edition. New York: Cambridge University Press.

4

Post-Occupancy Evaluation Processes in Six Federal Agencies

The federal government is the largest owner of facilities in the United States. More than 30 individual federal agencies own, use, and acquire facilities to support agency missions and programs. Some federal agencies conduct post-occupancy evaluation (POE) and lessons-learned programs as ways to improve customer satisfaction, to increase building quality and performance, and to facilitate organizational learning. This chapter provides information about POE processes in six federal agencies: the U.S. Air Force, Office of the Civil Engineer; the General Services Administration, Public Buildings Service (PBS); the Department of the Interior, National Park Service (NPS); the U.S. Navy, Naval Facilities Engineering Command (NAVFAC); the U.S. Department of State, Office of Overseas Buildings Operations (OBO); and the U.S. Postal Service (USPS).

The six agencies are sponsors of the Federal Facilities Council and volunteered to participate in the study. The information was gathered through a questionnaire and telephone interviews conducted by Krista Waitz of Kwaitz Consulting. National Research Council staff wrote the summary of findings and descriptions of POE programs.

The study design was not a scientific one, nor was it based on random sampling. Thus, the information provided should not be generalized. The remainder of this chapter contains information about the survey questions, a summary of findings, and descriptions of the POE programs in the six agencies.

SURVEY QUESTIONS

In January 2001, a questionnaire was designed and issued to six sponsor agencies of the Federal Facilities

Council who volunteered to provide information about their post-occupancy evaluation processes. In some cases, more than one person in the agency responded and the responses were combined and reconciled. The agency representatives were asked to respond to the following questions:

1. Approximately what year did your agency establish a post-occupancy evaluation program? On average, approximately how many POEs have been conducted by your agency in each of the last five fiscal years?
2. What were the driving factors for establishing a POE program?
3. What is the focus of the POE information-gathering process (e.g., user satisfaction, achievement of design objectives, building performance, other)?
4. What are the expectations, goals, and objectives for the program? Have they been achieved? Have there been unanticipated results?
5. How is the information gathered through POEs tied into the feedback loop (lessons learned) for planning, programming, and capital asset management?
6. Is the information gathered through POEs used in real estate decision-making and capital asset management? If yes, please note what information is used, how, and when it is used. If no, why not?
7. Is it your agency's policy to conduct POEs for all buildings or for selected facilities? What are the criteria for determining whether a POE will be conducted?

8. What do you consider to be barriers to conducting successful POEs?
9. Who is typically involved in conducting a POE in terms of in-house personnel and consultants? (Please list positions or types of skills involved not individuals.) What types of technologies are used?
10. What is the estimated cost in time and dollars for conducting a typical POE?
11. What data collection methods, technologies, and survey forms have been used over the life of the program? Please provide copies of survey forms that have been used.
12. To what extent does your agency make use of data management systems, Internet tools, or other information technology applications to share information and disseminate results of POEs?

SUMMARY OF FINDINGS

Establishment of POE Program, Timing of Surveys, and Number Conducted

Each of the six agencies studied had had a POE program in place at least since the 1980s. The POE programs of the PBS, NPS, and NAVFAC were being restructured to meet new objectives, and the results of the reorganized programs were not yet available.

The number of POEs conducted annually, on average, ranged from less than 1 to 30. Post-occupancy evaluations are typically performed within 4 to 24 months following occupancy of a new or renovated facility and are performed only once for an individual building.

Focus of POE Programs

Each of the six agencies used POEs to determine client or user satisfaction at some level, but it also used them to fulfill other objectives. These objectives included determining building defects within the construction warranty period, supporting design and construction criteria, supporting performance measures for asset management, evaluating construction inspectors, lowering facility life-cycle costs by identifying design errors that could lead to increased maintenance and operation costs, clarifying design objectives, improving building performance, and supporting corporate sales and image objectives. The restructured programs in NAVFAC and PBS are focused on developing metrics for client satisfaction and for management-related issues.

POE Process and Technologies

No two agencies use the same process or tools for conducting POEs and capturing lessons, although some share common elements. The National Park Service is in the process of developing new procedures and tools for conducting POEs and sharing lessons learned. Currently, at the NAVFAC, an independent agent remote from the designer of record conducts the POE using a statistically based questionnaire; a focus group discussion is then conducted to summarize the results of the survey. The NAVFAC criteria office administers survey results. The questionnaire is being modified so that it can be administered from field agencies and can be completed on-line or downloaded from the Web. The survey documents and results can be downloaded from a database on the Web.

The Air Force, in contrast, uses a questionnaire administered by a staff team; feedback is given primarily to the construction agent, although the Air Force plans to also share the results with users, the base civil engineer, and the major command.

The Office of Overseas Buildings Operations uses pre-trip user questionnaires, on-site interviews, and facilitated town meetings to gather the information, which is then summarized in a formal report. OBO's POE is conducted by a multidisciplinary in-house team that is customized to address known deficiencies.

The PBS performance measures-oriented approach uses a set of questionnaires developed in cooperation with the Center for the Built Environment at the University of California at Berkeley. The questionnaires are being designed to be administered over the Web. The lessons learned will provide input to the ongoing performance measures program of the Office of Business Performance.

The U.S. Postal Service uses two levels of surveys. The first, a basic questionnaire that can be completed in about an hour by the administrative service officer and the postmaster of a new facility, is required for all new construction. For larger, more complex projects, POEs are conducted over two to four days using a multidisciplinary team. Both types of surveys use electronic questionnaires in Microsoft Excel. The information gathered is sent directly to the staff maintaining the agency's design standards.

How POE Information Is Used

Information obtained from POE programs has been used by the OBO, NAVFAC, PBS, and USPS in support of their design criteria and guidelines. The NPS and the Air Force programs anticipate using POE results for the support of design criteria, among other objectives. To date, none of the six agencies reported that POE information was used directly in future real estate decision-making and capital asset management, although PBS's POE program was being restructured with those objectives in mind.

Barriers

A number of barriers to more effective use of POEs and lessons-learned programs were identified. These could be categorized generally as resources, feedback, and participation and commitment.

Resources

Several agencies noted it was difficult to obtain or earmark the funding needed to conduct POEs regardless of whether the POEs were to be conducted using consultants or in-house staff. In some cases, in-house staff may not be available to conduct the POE or may not have the technical skills needed for quality results.

Feedback

Because POEs often focus on identifying deficiencies, they risk becoming instruments to focus or deflect blame for unsatisfactory results. One agency cited the concern by senior executives that lessons learned may be considered a weakness by Congress or the Inspector General. Other agencies noted that conducting a focus group to achieve consensus about the cause of failures without judgmental discussion can be difficult, and as a consequence, they may be reluctant to do so.

Participation and Commitment

One agency noted that because of construction schedule constraints, staffs may be focused on future projects or those under visible construction. Thus, once a project has been completed and occupied, items such as financial closeout, construction as-builts, and POEs may not be a high-priority item and may not receive adequate oversight or attention. For programs adminis-tered through a headquarters' office, there may be a lack of field-level attentiveness to the process. Obtaining the clients' attention to ensure adequate participation in the survey or getting people not originally involved with the project to participate in a survey was also identified as a barrier. Organizational structures can also create barriers when responsibilities are assigned such that POE administration and database development require interoffice collaboration.

Costs

The costs reported to conduct POEs ranged from $1,800 for a simple standard questionnaire that could be completed in one hour to $90,000 for an in-depth analysis including several days of interviews, multi-disciplinary teams, site visits, and writing up reports. The costs did not include implementation of any changes resulting from a POE study. Costs per square foot of space evaluated were not available. Other variables accounting for the range of costs included whether the facilities were located in the United States or abroad, whether in-house staff or consultants were used, and how the resulting information was packaged and distributed.

DESCRIPTIONS OF POE PROGRAMS

National Park Service, Department of the Interior

In the mid-1980s the National Park Service completed formulation and development of an extensive POE program; however, due to changes in staff and downsizing of the NPS central design office, the program was not fully executed. In 1998, the POE program was reinstituted as part of the business practices for the NPS central design office. At that time, it was decided to reestablish the POE program for three reasons: (1) to create a feedback loop that would allow designers to interact with facility users and learn if facilities were meeting the needs of the users; it also served as an opportunity for users to assess their original requirements and determine if they had adequately identified their needs; (2) to evaluate the effectiveness of consultant construction inspectors who had recently replaced all NPS construction inspectors; and (3) to improve long-term facility life-cycle ratio costs by identifying any design errors that could lead to increased maintenance or operational costs.

The focus of the NPS POE program is to improve user satisfaction, building performance, and designer efficiency. The central design offices for the NPS seek to retain highly talented designers over a long-term career. By building relationships with facility users, designers better understand user needs and anticipate requests, allowing for more efficient use of design funds. The POE program seeks to build on this type of relationship through open communications and on-site review of completed construction projects. The NPS expects the value-added component of the POE will be in lessons learned and improved design efficiency. The NPS noted that maintenance and operations costs for facilities are escalating annually, and if the POE provides information for future designs that lead to improved maintenance or operations, the payback will be dramatic. In addition, if the POE provides feedback on products or techniques that improve user satisfaction, reduce maintenance, or improve operations, this information can be shared with designers to reduce the cost of design development.

Data collection tools and the forum for sharing information are still under development. However, the POE program is envisioned to include an evaluation form and follow-up meetings approximately six months to one year after completion of construction. The POE meetings will include the users, project manager, and design team captain. After completing the evaluation, the materials will be shared within the project management and design divisions. The results of the POE will also be placed in the central technical information files and may be accessed by various levels of project managers, designers, and other technical staff.

The NPS anticipates its POE teams will consist of (1) users, including park superintendent, chief of maintenance, park rangers, park interpreters, and administrative staff; (2) project managers, a multidisciplinary group consisting of park planners, architects, landscape architects, and various disciplines of engineers—the technical expertise will vary by project; and (3) design team members. Generally a team captain for a building project would be a senior architect; for a road construction project, a senior civil engineer or senior landscape architect would attend the POE; and for a utility project, a senior civil engineer would attend. Depending on the complexity of the project, other members attending a POE meeting could include a mechanical engineer or an electrical engineer.

Naval Facilities Engineering Command, Department of the Navy

The Naval Facilities Engineering Command has conducted POEs since the 1960s. In its original form, NAVFAC's POE program had an instructional base and was conducted by a project team after construction to identify lessons learned. In 1997, a new statistically based concept was initiated. The goal is to establish a statistical basis from which NAVFAC can measure agency improvement and work toward continuous improvement by implementing process changes and modifying design criteria. The POE program is intended to apply to each completed facility within its warranty period. On average, 20 POEs have been conducted in each of the last five fiscal years.

There were two driving factors for restructuring the POE program at NAVFAC. The first was a published Department of Defense-level survey of occupant satisfaction with their facilities. The second was a NAVFAC headquarters' management initiative to create metrics measuring how, or if, the agency exceeded client expectations. The agency focuses on improving client satisfaction, determining where its product lines or processes give rise to client dissatisfaction and improving the individual facility on which the survey is conducted. NAVFAC is issuing policy that will require a survey of all facilities within 6-10 months of building occupancy. The focus of the POE information-gathering process is to measure user satisfaction from the perspectives of the building owner, the customers (student, family occupant, day care family member, etc.), the building occupant at the working level, and the staff maintaining the facility. Client satisfaction at the user level is measured whether or not the client was involved in the planning, design, or construction phases. The measuring process includes a survey of participants after which a focus group discussion is conducted to summarize the positive and negative aspects of the facility.

Currently, all POE surveys are conducted by an "independent agent" remote from the designer of record, typically a consultant. The NAVFAC criteria office administers the surveys. The criteria manager for the facility type identifies criteria issues and action(s) to be initiated and implements interim guidance to the organization when required. The survey will be used in developing and modifying planning and design criteria. The data will be reviewed annually to determine process improvement needs. NAVFAC is working to determine how the survey will interface with its knowledge management system, combining to yield a single lessons-learned concept for the organization.

When the POE program process began, NAVFAC used a checklist format focusing only on design and construction. The survey was changed to accommodate all of the agency's processes (i.e., planning, design, construction, and maintenance turnover). Additional minor changes are anticipated to better assess safety and procurement issues. A copy of the survey is contained in Appendix D.

The database is being modified to Web-enable the survey content and to create a field-managed site as opposed to a headquarters central database. In addition, the database is being made integral to corporate ORACLE-based management systems. It will draw information from the management system and alert assigned individuals when it is time for the survey.

Office of the Civil Engineer, U.S. Air Force

The focus of the Air Force POE information-gathering process is user satisfaction, achievement of design objectives, and improved building performance. The purpose of the POE is to note all defective work, report construction deficiencies to the construction agent for correction by the contractor, and document problems or mistakes made during design for use as lessons learned on similar projects.

The POE is conducted by a staff team using a questionnaire. The Air Force plans to share the results of the POE with the construction agent, the user, the base civil engineer, and the major command for use in any future designs and for incorporation into Air Force design standards. It is Air Force policy for a POE to be accomplished sometime during the ninth to eleventh month following beneficial occupancy (acceptance of the facility by the user agency).

Office of Overseas Buildings Operations, Department of State

At the Office of Overseas Buildings Operations (formerly Foreign Buildings Operations) of the U.S. Department of State, lessons learned and design or construction alerts have been issued since 1985. The driving factors for establishing a POE program were the concern for user satisfaction, comfort, and safety and a general desire to capture best practices. The focus of the POE information-gathering process is on user satisfaction, achievement of design objectives, and building performance, including interior flexibility and functionality.

The POE methodology followed was adapted from the U.S. Postal Service. Over the life of the POE program, occupant surveys, on-site interviews, and facilitated town meetings have been used for data collection. The first step in the POE process is to send an occupant survey to an overseas post. The preliminary results from the occupant survey are used to determine what disciplines should be represented on the multidisciplinary team that will be conducting the POE; thus, the team is customized to address known deficiencies. Typically, architects; electrical, mechanical, and structural engineers; facility maintenance; and security specialists are involved in conducting POEs. Once at the site, the team conducts a walk-through followed up by interviews with occupants. At the conclusion of the site visit, members of the team reconvene to discuss their observations and to generate recommendations for a report. Due to constrained resources, a POE may consist solely of the occupant survey.

One result of the POE program has been the revision of design guidance on such topics as roofs, elevators, and Ambassadors' residences. POE results were also used for developing a serviceability demand profile for generic embassy office buildings slated for design and in design guidelines for future embassies being acquired under a specialized procurement process.

Public Buildings Service, General Services Administration

The Public Buildings Service (PBS) of the General Services Administration (GSA) first instituted a POE program oriented toward design and construction criteria development in 1977. In-house technical experts, including an environmental psychologist, architects, and engineers, conducted the surveys. The program was curtailed in 1982 due to a reduction in staff. PBS' POE program was reinstituted in 1986, using contractor support. Between 1986 and 2000, the PBS completed approximately 30 POEs for a variety of projects, including courthouses, office buildings, U.S. border stations, major renovations, and historic restorations. In 2000, the PBS restructured its POE program to focus on performance measures for asset management. Both the design criteria-oriented POE program and the performance measure-oriented POE program were intended to provide an information stream that would inform program managers, criteria managers, and project managers about design- and delivery-related problems and associated best practices.

Criteria–Based POEs

Criteria-based POEs were developed to provide technology- and procedure-based feedback to those in PBS's central office responsible for national program and design criteria direction. Equally important were the perceived benefits of offering those same lessons to the delivery teams responsible for new projects. The primary focus was on building systems evaluation, client satisfaction surveys, and interviews with major client agencies. The building systems evaluations included functionality issues as well as an overview of how well the building complied with design criteria. The surveys were customized for different types of buildings and project delivery systems, including office buildings, courthouses, border stations, and lease-build or design-build projects. (An example is included in Appendix D.) Major lessons from the POE process related to long-term building maintainability, building functionality, client needs, and property manager and asset management needs. Specific issues concerned energy efficiency; indoor air quality; heating, ventilation, air conditioning, and electrical systems; thermal comfort; design of loading docks; access to equipment; window washing; and accessibility for the physically disabled. The criteria were constantly revised to incorporate lessons from POEs.

From 1986-2000, criteria-based POEs typically involved a team of five to six design-related disciplines, using outside architect-engineer professional services. PBS central office coordination and involvement were provided through site visits, access coordination, and report critiques. A report and a 30-minute video were created and distributed to each of GSA's field offices (11 regions) and to senior-level management at headquarters.

The tieback to lessons learned was direct: the people who conducted the POEs also developed the criteria. However, because the POE reports were voluminous and oriented toward detailed evaluations of technology applications, getting project managers and designers to read and adhere to findings was a challenge. Thus, various forms of information exchange media were applied to mitigate this problem, including condensed videotapes and interactive "lessons-learned" compact disks

(CDs). A compendium of lessons was distributed every three years. In 1998, PBS prepared a Compendium of Lessons Learned CD-ROM that was widely distributed, and a DVD (Lessons Learned, Volume 2) is being prepared for mass distribution. The POEs are also in the construction criteria database of the National Institute for Building Science.

A second program that resulted from the POE lessons is HVAC (heating, ventilation, air conditioning) Excellence in Federal Buildings. PBS held numerous workshops that included staff, architects, engineers, and representatives from professional societies and technical organizations as part of an awareness program to highlight HVAC issues.

Performance Measure-Oriented POEs

The driving factor behind the shift to a performance measure-oriented POE program was PBS senior management's desire to evaluate how well PBS's assets are achieving their objectives on a project and program basis. Performance measure-oriented POEs are being pursued to help indicate whether delivery practices and criteria are effective and to identify systemic problems, whereby specialty studies could be pursued.

The focus of this POE information-gathering process is financial asset assessment. A set of extensive questionnaires is used in an attempt to identify customer satisfaction with various building components or features. Different questionnaires are directed to different key personnel, including operating staff and design-delivery team members. The measures are intended to help determine if GSA is meeting a number of key management indicators including comparison of construction "pro forma" with final pro forma, maintenance and cleaning costs benchmarked against national standards, utility costs, sustainability, energy usage against FY 2010 goals, accessibility for the physically disabled, and client satisfaction.

Support for questionnaire development and database management is currently provided by the Center for the Built Environment, within the University of California at Berkeley. Questionnaire delivery and assessments are being coordinated by senior architects and engineers within the PBS Office of Business Performance. PBS is currently designing tenant, operations and maintenance, and design and construction survey tools that can be administered over the Web.

The goal is to perform a POE for every Congressionally approved new building one year after full occupancy. Once the performance measure-oriented POE program is fully implemented, 10-20 POEs will likely be completed each year. The tieback from the ongoing performance measure POE program will go from PBS's Office of Business Performance to other appropriate offices within the agency.

U.S. Postal Service

The U.S. Postal Service established its POE program in 1986. The driving force behind the establishment of the POE program was the desire to improve the planning, design, and construction of future facilities. The focus of the POE information-gathering process is user satisfaction (customers and employees), clarification of design objectives, achievement of design objectives, building performance relative to technical systems such as cooling and lighting, and supporting corporate sales and image objectives and economics. On average, the agency has conducted approximately 30 POEs in each of last five fiscal years.

The first POE application was an effort to standardize the hundreds of Postal Service retail spaces produced each year. Appropriate design was found to be a much more powerful factor in customer satisfaction than had been anticipated. Also, building image was a much more powerful support for overall corporate identity than previously realized.

Two levels of POEs are used currently. A basic POE (completing the questionnaire) is required for all new construction and for owned facilities greater than 9,000 square feet within four and six months of occupancy. The administrative service office manager and the postmaster complete the basic POE questionnaire, which typically takes 30 minutes to an hour. An example is included in Appendix D.

More extensive POEs are conducted on larger projects (more than 30,000 square feet) or other special projects. These POEs involve architectural or engineering firms (including environmental psychologists as consultants) and are conducted over a period of two to four days. Customers and employees are interviewed, and extensive lighting and HVAC data are gathered.

The results of POEs go directly to the staff maintaining the Postal Service Building Design Standards. The information gathered through the POE process is not used in real estate decision-making or capital asset management; however it is used in planning, design, and construction decisions. Real estate decisions are affected only as site-planning criteria are modified.

5

Post-Occupancy Evaluations and Organizational Learning[1]

Craig Zimring, Ph.D., Georgia Institute of Technology
Thierry Rosenheck, Office of Overseas Buildings Operations, U.S. Department of State

Federal building delivery organizations face intense pressures. Not only must they provide buildings on time and within budget, but they have increased demands. They are called on to deliver buildings that are better: more sustainable, accessible, maintainable, responsive to customer needs, capable of improving customer productivity, and safer. In many cases, they must achieve these goals with fewer staff.

Some organizations have faced these pressures proactively, by creating formal processes and cultural changes that make their own organizational learning more effective. In this chapter we adopt the approach to *organizational learning* of Argyris (1992a), Huber (1991) and others. We mean that organizations are able to constantly improve the ways in which they operate under routine conditions, and they are able to respond to change quickly and effectively when needed (Argyris, 1992a). Learning is "organizational" if it is about the core mission of the organization and is infused through the organization rather than residing in a few individuals. More simply, in the words of Dennis Dunne, chief deputy director for California's Department of General Services, they "get it right the second or third time rather than the seventh or eighth." By being more systematic about assessing the impact of decisions and being able to use this assessment in future decision-making, building delivery organizations are able to reduce the time and cost to deliver buildings and increase their quality.

Some of the best models come from private sector organizations. For example, Disney evaluates everything it does and has been doing so since the 1970s. Disney has at least three evaluation programs and three corresponding databases: (1) Disney tracks the performance of materials and equipment and records the findings in a technical database. (2) Guest services staff members interview guests about facilities and services, focusing on predictors of Disney's key business driver: the intention of the customer to return. (3) A 40-person industrial engineering team conducts continuous research aimed at refining programming guidelines and rules of thumb. The industrial engineering team explores optimal conditions: What is the visitor flow for a given street width when Main Street feels pleasantly crowded but not oppressive? When are gift shops most productive? This research allows Disney to make direct links between "inputs" such as the proposed number of people entering the gates and "outputs" such as the width of Main Street.

The Disney databases are not formally linked together but are used extensively during design and renovation projects. They have been so effective that the senior industrial engineer works as a coequal with the "Imagineering" project manager during the programming of major new projects.

Disney is a rare example. It uses an evaluation program to do the key processes that organizational learning theorists argue are key to organizational learning (Huber, 1991):

1. monitoring changes in the internal and external business environment,

[1]For their generous and thoughtful input we would like to thank Stephan Castellanos, Dennis Dunne, Gerald Thacker, Lynda Stanley, Polly Welch and Richard Wener.

2. establishing performance goals based on internal and external influences,
3. assessing performance,
4. interpreting and discussing the implications of results,
5. consolidating results into an organizational memory,
6. widely distributing findings and conclusions,
7. creating a culture that allows the organization to take action on the results,
8. taking action based on organizational learning.

Post-occupancy evaluation (POE) practice focuses mostly on individual project support and analysis rather than on lessons-learned. Although POE potentially provides a methodology for all of these processes, POE practice has historically had a more narrow focus on assessing performance and interpreting results. POE has often been used as a methodology aimed at assessing specific cases, while the other processes are seen as part of strategic business planning. Even when evaluators have been able to create databases of findings, they have often been used to benchmark single cases rather than to develop more general conclusions.

Structured organizational learning is difficult. It requires the will to collect data about performance and the time to interpret and draw conclusions from the data. More fundamentally, learning involves risk and change. Learning exposes mistakes that allow improvement but most organizations do not reward exposing shortcomings. Learning brings change and organizations are usually better at trying to ensure stability than at supporting change.

In this chapter, we explore how a variety of public agencies and some private ones have used POE successfully for organizational learning. We discuss the "lessons-learned" role of evaluation rather than the project support and analysis role. We have examined written materials from 18 POE programs and interviewed participants wherever possible. We explored whether POE-based organizational learning appeared to be going on, whether the organizations had established support for learning, and the nature of the learning.

Did POE-Enabled Organizational Learning Occur?

For example, in looking for evidence of organizational learning we asked the following questions:

- Are participants in building projects, including internal project managers, consultants, and clients, aware of POEs or POE results, either from personal participation or from written results?
- If so, were POE results consciously used in decision-making about buildings? For example, are they used for programming, planning, design, construction, and facilities management?
- Can we see evidence that POE results are part of reflection and discussions about how to do a good job, among peers and with supervisors?
- Are POE results consciously used to refine *processes* for delivering buildings in terms of either formal process reflected in manuals or informal rules of thumb and customs?
- Are people who make *policy* about buildings, such as policy directives, design guidelines, and specifications, aware of POEs?
- If so, is POE explicitly used in formulating policy?

Were the Conditions for Organizational Learning Present?

As we attempted to document organizational learning we were trying to understand the conditions that foster or thwart it:

- Does the organization have an infrastructure for learning? For example, are results from POEs consolidated in some way, such as in reports or databases? Is this consolidated information distributed, either internally or to consultants or the public?
- Is there a mechanism for ensuring that this information is kept current?
- If lessons are made available, do they support the kinds of decisions that are made by the organization? Are they likely to seem authentic and important to decision-makers? Are the implications of results made clear, or do busy decision-makers need to make translations between results and their needs?
- Are there incentives for accessing the data, using the results, and contributing to the lessons-learned knowledge base? For example, are internal staff or consultants evaluated on use of lessons-learned? Are they rewarded in some way for participation? Are consultants rewarded for partici-

pation? Are they rewarded for good performance as judged by the POE?

- Are there *disincentives* for participating in lessons-learned programs? If an innovative initiative receives a negative evaluation, is it treated as an opportunity for organizational learning or as a personal failure?
- Is there a perception of high-level support for the lessons-learned program? Many organizations create frequent new initiatives, and seasoned staff often perceive these as the management "infatuation du jour": wait a day and it will change.

How Can Organizations Develop Useful Learning Content?

We also assessed the *content* of the organizational knowledge. We asked the following:

- Has the organization produced a shared view of what makes a good building, in terms of either process or product? For example, has the organization been clear about key design and programming decisions and about how these decisions link to the client's needs? Are these contributed to by POE?
- Has the organization created an organizational memory of significant precedents? Are these precedents described, analyzed, or evaluated in meaningful ways?
- Is this view tested and refined through POE or similar processes?

In this chapter, we briefly report our findings and analysis. We discuss four topics:

1. What is post-occupancy evaluation? What is its history, and how has this contributed both to its potential for and difficulties in achieving organizational learning?
2. Do organizations do POE-enabled organizational learning?
3. How have organizations created the appropriate conditions for learning through POE?
4. How have they created a knowledge base for building delivery and management?

BRIEF INTRODUCTION TO POST-OCCUPANCY EVALUATION

Post-occupancy evaluation grew out of the extraordinary confluence of interests among social scientists, designers, and planners in the 1960s and 1970s (see, for example, Friedmann et al., 1978; Shibley, 1982; Preiser, et al., 1988). Early POE researchers were strongly interested in understanding the experience of building users and in representing the "nonpaying" client (Zeisel, 1975). Many early POEs were conducted by academics focusing on settings that were accessible to them, such as housing, college dormitories, and residential institutions (Preiser, 1994).

During the 1980s, many large public agencies developed more formal processes to manage information and decisions in their building delivery processes. As planning, facilities programming, design review, and value engineering became more structured, agencies such as Public Works Canada and the U.S. Postal Service added building evaluation as a further step in gathering and managing information about buildings (Kantrowitz and Farbstein, 1996).

This growth of POE occurred while politicians and policy analysts were advocating the evaluation of public programs more generally. Campbell and many others had been arguing at least since the 1960s that public programs could be treated as social experiments and that rational, technical means could contribute to, or even replace, messier political decision-making (Campbell, 1999). A similar argument was applied to POE: statements of expected performance could be viewed as hypotheses that POE could test (Preiser et al., 1988).

The term *post-occupancy* evaluation was intended to reflect that assessment takes place after the client had taken occupancy of a building; this was in direct contrast to some design competitions where completed buildings were disqualified from consideration or to other kinds of assessment such as "value engineering" that reviewed plans before construction. Some early descriptions focused on POE as a stand-alone practice aimed at understanding building performance from the users' perspectives. Some methodologists have advocated the development of different kinds of POEs, with different levels of activity and resource requirements (Friedmann et al., 1978; Preiser et al., 1988). For example, Preiser advocated three levels of POE: brief indicative studies; more detailed investigative POEs; and diagnostic studies aimed at correlating environmental measures with subjective user responses (Preiser, 1994). Whereas there was little agreement about specific methods and goals, most early POEs focused on systematically assessing human response to buildings and other designed spaces, using methods such as questionnaires, interviews, and observation, and sometimes linking these to physical assessment (Zimring, 1988).

Over the years, many theorists and practitioners have grown uncomfortable with the term POE; it seems to emphasize evaluation done at a single point in the process. Friedmann et al. (1978) proposed the term "environmental design evaluation." Other researchers and practitioners have suggested terms such as "environmental audits" or "building-in-use assessment" (Vischer, 1996). More recently, "building evaluation" and "building performance evaluation" have been proposed (Baird et al., 1996). Nonetheless, for historical reasons the term post-occupancy evaluation remains common, and we use it in this chapter for clarity.

Other discussions of evaluation emphasized the importance of embedding POE in a broader program of user-based programming, discussion, and design guide development, proposing terms such as "pre-occupancy evaluation" (Bechtel, 2000), "process architecture" (Horgen et al., 1999), and "placemaking" (Schneekloth and Shibley, 1995). As early as the 1970s, the Army Corps of Engineers conducted an ambitious program of user-based programming and evaluation that resulted in some 19 design guides for facilities ranging from drama and music centers to barracks and military police stations (Schneekloth and Shibley, 1995; Shibley, 1982, 1985). More recently, POE has been seen as part of a spectrum of practices aimed at understanding design criteria, predicting the effectiveness of emerging designs, reviewing completed designs, and supporting building activation and facilities management (Preiser and Schramm, 1997). With growing concerns about health and sustainability, several programs have also linked user response to the physical performance of buildings, such as energy performance (Bordass and Leaman, 1997; Cohen et al., 1996; Leaman et al., 1995) or indoor air quality (Raw, 1995, 2001).

POE methodologists and practitioners have identified several potential benefits of POE (Friedmann et al., 1978; McLaughlin, 1997; Preiser et al., 1988; Zimring, 1981):

- A POE aids communications among stakeholders such as designers, clients, end users, and others.

- It creates mechanisms for quality monitoring, similar to using student testing to identify underperforming schools, where decision-makers are notified when a building does not reach a given standard.
- It supports fine-tuning, settling-in, and renovation of existing settings.
- It provides data that inform specific future decisions.
- It supports the improvement of building delivery and facility management processes.
- It supports development of policy as reflected in design and planning guides.
- It accelerates organizational learning by allowing decision-makers to build on successes and not repeat failures.

This chapter focuses primarily on the use of POE for improving organizational learning.

DO ORGANIZATIONS DO POE-ENABLED ORGANIZATIONAL LEARNING?

As discussed above, we reviewed materials from some 18 organizations that are currently doing POEs or have done so in the past. In looking at organizations that have active POE programs, we found that members of project teams, including project managers, consultants, and clients, tend not to be aware of POEs, unless a special evaluation has been conducted to address a problem that the team is facing. Where they are aware of the POEs, team members often do not have the reports from past POEs at hand and do not apparently use POE results in daily decision-making.

Mid-level staff tend to be more aware of POEs. In particular, staff responsible for developing guidelines and standards are often aware of POE results. For example, in the U.S. Postal Service, the staff who maintain guidelines also administer POEs; the POEs conducted by the Administrative Office of the U.S. Courts are used directly by the Judicial Conference to test and update the U.S. Courts Design Guide.

We were not able to find situations where senior management used POEs for strategic planning. POEs have the potential for supporting "double-loop learning" (Argyris and Schon, 1978)—that is, not only to evaluate how to achieve existing goals better but also to reflect on whether goals themselves need to be reconsidered. However, we were not able to find cases where this actually occurred.

We were not able to find many compilations of POE findings, although several organizations such as the U.S. Army Corps of Engineers, U.S. Postal Service, Administrative Office of the U.S. Courts, General Services Administration, and others have incorporated POEs into design guides. Disney and the U.S. Department of State have incorporated POE into databases of information. These are discussed in more detail below.

It does appear that POEs are not used for their full potential for organizational learning. In particular, we were not able to find many circumstances where POE was part of an active culture of testing decisions, learning from experience, and acting on that learning. There are two major reasons for this:

1. Learning is fragile and difficult, and many organizations have not created the appropriate conditions for learning. If learning is to be genuinely "organizational," a large number of staff must have the opportunity to participate and to reflect on the results in a way that enables them to incorporate the results into their own practice. Potential participants must see the value for themselves: there must be incentives for participating. Also, evaluation will sometimes reveal that building performance does not reach a desired standard. This is, of course, the value of POE, but many organizations punish people when innovations do not work. In addition, many organizations simply do not make information available in a format that is clear and useful to decision-makers.

2. Many organizations have not created a body of knowledge that is valuable in the sense that it provides a coherent, integrated body of knowledge that is helpful in everyday decision-making. Knowledge tends to be informal and individual.

WAYS TO CREATE THE APPROPRIATE CONDITIONS FOR LEARNING THROUGH POE

Create Broad Opportunities for Participation and Reflection

Our research suggests that POE-based knowledge is not widely shared within most organizations. One way to achieve this sharing is through direct participation in evaluations. Seeing how a facility works while hearing directly from users is a memorable experience. Also, the process of analyzing and writing up the results from

an evaluation can help decision-makers reflect on the implications of the results and make links to their own practice.

A group of evaluators in New Zealand developed a "touring interview" methodology to allow decision-makers to actively participate in evaluations with little training and only modest commitment of time (Kernohan et al., 1992; Watson, 1996; 1997). For example, in an active evaluation program with more than 80 completed evaluations, Bill Watson, a consultant to public and private clients, takes building users on a tour of the building and asks open-ended questions—for example, "What works here?"—as well as more specific probes about the functions of spaces and systems. POE reports are mostly verbatim comments by users and are sorted into categories such as "action for this building" or "change in guidelines for future buildings." This approach is quite inexpensive and can be completed with several person-days of effort. The experience is vivid for the participants and produces results that are imageable and articulate. It also allows participants to discuss relative priorities and values. However, because each touring interview group varies, it is more difficult to compare evaluations of different settings.

This kind of participatory evaluation can be an extension of existing processes for receiving feedback from customers. Project managers in Santa Clara County, California, were tired of receiving a storm of requests from users after they moved into a building. These were difficult to direct to contractors, suppliers, and others. They contracted with consultants Cheryl Fuller and Craig Zimring to create a Quick Response Survey (QRS) aimed at organizing and prioritizing user needs about three months after buildings were occupied. All building users fill out a one-page questionnaire, and project managers follow up with a half-day walk-through interview of the building with the facility manager and staff representatives. The project managers then prioritize requests and meet with the client organizations. The State of California Department of General Services is further developing the QRS and will have evaluators enter results into a lessons-learned database.

A lessons-learned program initiated in 1997 for New York City to examine the success of school projects in the state was aimed at participation by consultants. The School Construction Authority (SCA), whose membership is appointed by the governor, the mayor, and the New York City Board of Education, was charged with the program. To get the program approved, SCA, under the leadership of consultant Ralph Steinglass, adopted a simple methodology—require the architect or engineer of record to conduct the POE. The rationale was that this would guarantee that designers would confront how users responded to their designs and force a lessons-learned loop in the design process. About 20 POEs have been completed. To ensure reliability, SCA reviewed the results before approving the POEs. In some cases, the architects or engineers had to reschedule their interviews when they were suspected of introducing a bias or continue their investigation if they failed to include critical areas required in the study.

The three programs described above involve evaluation by the people who designed and managed the project. As such, participatory evaluation is well suited to supporting learning by in-house project managers and consultants. Whereas the New Zealand projects are led by consultants, the quick-response projects and the SCA projects are conducted entirely by the consultants or project managers.

Create Incentives for Participation

Most building professionals are interested in doing a good job and see value in POE. However, as personal time management consultant Stephen Covey has argued, things that are merely important often lose out to things that are urgent: general benefits such as long-term learning often lose out when professionals are faced with the pressing matters of everyday life. When more specific incentives are offered, it often increases participation in a POE program.

The drug company Ciba-Geigy has used direct monetary incentives. The architectural and engineering firm HLW and the contractor Sordoni Skanska Construction put their design and construction profits ($300,000 and $1.2 million, respectively) at risk based on performance on schedule, cost, and user satisfaction for the new $39 million Ciba-Geigy Corporation's Martin Dexter Laboratory in Tarrytown, New York. One-third of the profits was based on user satisfaction responses to 14 survey questions: heating, ventilation, air conditioning, acoustics, odor control, vibration, lighting, fume-hood performance, quality of construction (finishes), building appearance, and user-friendliness. The questions were binary-choice (acceptable-not acceptable), and the building had to reach 70 percent satisfaction to pass the test. Some aspects such as sound transmission were

also assessed using physical measures; if the user satisfaction measures did not reach criterion, physical measures could be substituted (Gregerson, 1997). The designers and contractors consulted the scientists throughout the process, showing them alternatives for the façade design and full-scale mockups of the range hoods. The building passed on all criteria except satisfaction with the range hoods, which were modified after the evaluation as a response to user input. Whereas some aspects of this testing process might be questionable—Should you evaluate an entire building on 14 yes-no scales? Should maintenance and operating experience be included?—this process gained from participation throughout. The design firm, contractors, management, and scientists all participated in establishing the criteria at the outset, and the financial incentive encouraged the contractor and designers to consult the users at every step in the process. It is difficult to document the learning benefit of this process, but the contractor, Sordoni Skanska, has since used POE-based incentives in several other projects and has refined the way in which buildings are delivered.

The California Department of General Services is planning to include the results of POEs as part of the review of qualifications when selecting consultants and contractors. This has strongly increased the interest in POEs by participating firms. We are unaware of any POE programs that provide incentives for internal staff members to participate in evaluations, though several programs have discussed such incentives, such as providing a free vacation day as a reward for adding data to the knowledge base or providing a mini-sabbatical for participating in evaluations or a lessons-learned program. Disney provides a powerful, if indirect, incentive: knowledge. Only the industrial engineers have access to key data and they then become valuable members of the design team.

Reduce Disincentives: Create Protected Opportunities for Innovation and Evaluation

Organizational learning consultants have long pointed to an inherent contradiction in many organizations. Whereas most organizations espouse innovation and learning, they behave in ways that limit it. We recently participated in a meeting where an organization had used an innovative building delivery strategy with which it was not familiar. They had left out a key review step. When this became clear, a senior manager turned to the project manager and said: "We would have expected someone at your level to do better." The message to everyone in the room was clear: avoid innovation and avoid evaluation! This syndrome—focusing on the individual rather than the performance, blaming the innovator rather than learning from the innovation—is pervasive among organizations more generally (Argyris, 1992b; Argyris and Schon, 1978). However, some building-delivery organizations have used POE to at least partially overcome it.

Some organizations have done this by explicitly sanctioning "research" with the attendant acknowledgment that innovations might not succeed. For example, the General Services Administration's (GSA) Public Buildings Service has recently appointed a director of research. The first director, Kevin Kampschroer, has a budget to conduct, synthesize, and distribute research, including POE. The use of the term "research" carries with it the understanding that not all efforts are successful, and the budget provides some time for reflection about findings. To date, much of the research is conducted by academic consultants who bring outside learning into GSA. However, GSA is also looking at ways to broaden internal ownership of the research program.

GSA has also created an active "officing" laboratory in its own headquarters' building. The lab, supervised by Kampschroer, is one floor of actual workspace that includes an innovative raised floor heating, ventilating, and air-conditioning system and several brands of modular office furniture systems. It also explores design to support teamwork, with many small conference rooms and meeting areas. The workers are frequently surveyed and observed, and the lab also becomes a place where clients can see alternative office layouts.

The U.S. Courts and the General Services Administration Courthouse Management Group are considering developing a different kind of laboratory: a full-scale courtroom mockup facility where new courtroom layouts and technologies can be tested and refined at relatively low cost and risk. This facility, to be constructed at the Georgia Institute of Technology, would allow mock trials to be conducted and would provide training for judges, staff, and lawyers.

Another way to reduce the personal and organizational cost of experimentation is by starting small with projects that have an experimental component. The innovation can be evaluated and considered for broader adoption. For example the U.S. Department of State Office of Overseas Buildings Operations (OBO) tries

out innovations on a limited number of projects before rolling out the innovation to the larger organization. This office has recently used building serviceability tools and methods (Davis and Szigeti, 1996) for programming and design review for the new embassies in Dar es Salaam and Nairobi.

The State of Minnesota Department of Natural Resources has used POE to evaluate two innovative regional centers. In each of these cases the organizational learning effort provided some additional resources for data collection and reflection as well as the clear designation that this was an innovative effort that might not be fully successful.

In many organizations it is risky to be the first one to try an innovation. Massachusetts Institute of Technology organizational consultant Edgar Schein has proposed that while organizations may benefit greatly from consultants, they often find the experience of peers more helpful when they actually move to implementing an innovation. Schein has called for "learning consortia" where people can get advice from peers in other organizations and learn from their experience (Schein, 1995). He argues that although such learning consortia may be effective at all levels of an organization, they are particularly effective among chief executive officers (CEOs) or upper to mid- level managers. Although the prototype, laboratory, and learning consortium efforts are quite different, all reduce the disincentives for innovation and evaluation by allowing innovation and evaluation at relatively low personal and organizational cost.

Provide Access to Knowledge for Different Audiences

The simplest barrier to using POE for organizational learning is when POE results are not available to decision-makers. Many organizations produce POEs as case study reports that are not widely distributed. Part of this may be due to the history of POE, which has focused on single case studies, and part may be because of the perceived disincentives to distributing information that might be seen as critical of internal efforts or individuals. Part of the problem is the simple technical difficulty of distributing printed information, and this has become a lot easier with the Internet and intranet and virtual private networks. The National Aeronautics and Space Administration makes its lessons learned database available to all authorized staff and contractors. In the United Kingdom, Adrian Leaman and Bill Bordass have created an interactive Web site for the 18 buildings they have evaluated as part of the PROBE project. Funded by the Building Research Establishment and Building Services Journal, PROBE stands for post-occupancy review of buildings and their engineering.

Some organizations have overcome some of these issues by creating design guides and databases of POE information. Agencies such as the Administrative Office of the U.S. Courts, the U.S. Postal Service, and the General Services Administration have created design guides that are widely distributed.

As we have suggested, the problem with organizational learning is only partially technical. The tools for creating Web sites and databases are now widely available and inexpensive. A useful Web site requires the initiative to collect the information, the time to make sense of it, and the will to share it.

Part of the problem with building delivery organizations and design projects is that they represent many different professional cultures. Engineers tend to take a technical problem-solving approach. Architects are often interested in form. Clients might be interested in the usability and experience of the building. Senior managers might be searching for help in setting strategic directions, whereas project managers might be interested in lessons learned about specific materials or equipment. Part of the challenge in creating any database or report is translating between these different professional cultures, and evaluators have not always been successful at doing this.

Reduce Uncertainty by Upper Management's Commitment

Participants in POE programs report that uncertainty about senior management's commitment to the program is a key disincentive to participation. Sometimes the lack of commitment is seen in lack of resources, but it is often manifest in lack of visible endorsement for the program and lack of commitment to the two- to five-year time span necessary to see results in terms of organizational learning. Savvy staff have learned not to genuinely commit to the management's infatuation du jour, knowing that it will change quickly.

CREATING A KNOWLEDGE BASE FOR BUILDING DELIVERY AND MANAGEMENT

Most fundamentally, organizational learning for a building delivery organization is producing better

buildings more effectively. Given the large number of POEs that have been completed—longtime researcher and *Environment and Behavior Journal* editor Robert Bechtel estimates that more than 50,000 have been completed—one would expect that there would have been many books and guides that synthesize the results of POEs and tie them into a coherent guide for key programming and design decisions. However, such guides are relatively rare. In part, this is because of the focus of POE researchers and consultants on case studies. Much knowledge about buildings has been built up incrementally through negotiation on individual projects and programs, but organizations seldom take the time to identify the key strategic decisions that most affect their clients. Efforts such as American Society for Testing and Materials (ASTM) Building Serviceability Tools and Methods have begun to do this for programming and portfolio management, but we have seldom done this for POE.

In this section we examine several strategies that have proven successful for beginning to create this kind of knowledge base. In several cases, organizations have built on POEs that have been initiated for different purposes (or seem to us to be able to do so reasonably easily).

POE can be particularly successful in organizational learning if it links strategic facilities decisions to the "key business drivers" of the client organization. In the 1970s, the U.S. Army was shifting to an all-volunteer army. Potential recruits said that the aging facilities were a significant impediment to recruiting and retention, and the Army sought to renovate or rebuild many of its buildings. To help guide the multibillion dollar investment, the Army Corps of Engineers created a large program of participatory programming and evaluation, resulting in some 19 design guides (Shibley, 1985).

In the 1980s, the newly reorganized U.S. Postal Service (USPS) was losing customers to private competitors such as FedEx and UPS (Kantrowitz and Farbstein, 1996). Focusing initially on the customer experience with lobbies, the USPS contracted with Min Kantrowitz and Jay Farbstein and Associates to conduct focus group evaluations. This has led to a large and continuing program of evaluations and design guide development. New concepts of post office design are developed such as the retail-focused "postal store," innovative projects are designed, the projects are evaluated, and the ideas are refined and then incorporated into design guides. This program has sustained an ongoing process of testing and refining the design guides through evaluation and experience. More recently, the USPS has de-emphasized on-site evaluations. Most POEs now involve having facility managers fill out relatively brief mail-out surveys. The POE manager has found that the open-ended responses to the questionnaire are often most valuable in refining the USPS design guidelines because they are more specific than the scaled satisfaction responses.

BUILDING ON EXISTING EVALUATIONS

An organization can begin to rationalize its knowledge base by building on evaluations that occur for other reasons. As the experienced evaluator Bob Shibley has said, evaluations are easiest to justify if they bring project support, analysis benefits, and lessons-learned benefits (Shibley, 1985).

Building on Diagnoses of Troubled Settings

Sometimes a building is the subject of complaints or controversy; a POE can help to diagnose the source of problems and prioritize solutions. For example, the new San Francisco central library was a landmark when it opened in 1996, but it faced immediate controversy. Some of the initial programmatic assumptions continued to be debated—such as the wisdom of moving books to closed stacks to create room for computers. As a result, the mayor appointed an audit commission that recommended a POE, led by architect Cynthia Ripley and including the director of the Los Angeles library system. After interviewing staff and users, observing use, and analyzing records, the POE team highlighted problems with way-finding, flexibility, and public access to books. The POE recommended detailed renovations to reorganize the stacks and collection (Flagg, 1999; Ripley Architects, 2000). The POE is quite thorough in suggesting detailed specific changes, and the basis of these recommendations could potentially be turned into planning principles. Most significantly, this raises issues of programming process where the (former) library director went against the recommendations of his planning committee to reduce access to books in favor of closed stacks. It can support broader reflection about the role of libraries and physical structures in providing information in the age of computers.

Whereas the focus on understanding problems and failure provides a clear direction to POEs and has a

history in case studies of blast, earthquakes, and other building failures, these kinds of POEs carry special risks of becoming ways to focus (or deflect) blame.

Capitalizing on Evaluations of Innovations

Evaluation can help decide whether innovative buildings or building components should be considered for additional capital investment. For example, as mentioned above, the State of Minnesota Department of Natural Resources (DNR) has recently changed the way in which it manages the environment. Rather than organizing its staff by discipline, DNR now uses a matrix management system where decisions are made by a multidisciplinary group organized by ecosystem. The DNR is creating new regional centers that include wildlife biologists, air and water specialists, and others concerned with a given area. The centers are intended to encourage multidisciplinary collaboration and to be very "green." The DNR contracted with a university team led by Julia Robinson to evaluate two of the initial projects. The team made numerous recommendations. The POE was included as an appendix for the funding request for the third center. This was the first time in DNR's history that a capital request was fully funded on the first attempt, and the DNR was told that the POE was a major reason: it showed a high level of understanding of the project. This result provided an important incentive for DNR as an organization. However, the project also raised some issues about sustainability, and the internal staff did not feel that they had been fully consulted in the POE process. An additional team was hired to create design guidelines in close consultation with staff (M. Wallace, personal communication, 2000). Issues such as sustainability, which are undergoing rapid change, are particular candidates for "double-loop" learning where both goals and methods for achieving them are being developed, if appropriate conditions are established for discussion, reflection, and action.

Focus on "Learning Moments"

The Administrative Office of the U.S. Courts (AO) conducts a POE program that informs guidelines (in the U.S. Courts design guide). However, the AO has achieved organizational learning by linking the design guide to a strategic learning moment in the development of courthouses: the negotiation between judges and the building agent (the General Services Administration) about the scope and quality level for new court-houses. In the early 1990s, the U.S. government initiated the largest civilian construction program since the Second World War, projecting to spend more than $10 billion on 160 new courthouses. (The creation of new judgeships in the 1980s, concerns for increased security, and new technologies all necessitated new courthouses or major renovations.) However, both the judiciary and the GSA were being criticized by Congress for creating marble-clad "Taj Mahals." The AO initiated the POE program to identify necessary changes to the standards in the first edition of the design guide, to defend the judiciary against attack by documenting the efficacy of the design standards, and to inform the negotiation about issues such as the dimensions and materials of courtrooms and chambers. Information from POEs was also used in training workshops for judges and staff who were becoming involved in new courthouse design and construction. This program is run by the AO, but the design guide is actually created and vetted by a committee of the Judicial Conference, the group that sets broad policy within the federal judiciary. This program is quite unusual: it is the only case that we are aware of where a POE and design guide are developed by a client organization that does not build its own buildings.

The Minnesota Department of Natural Resources project also focuses on strategic moments, especially the approval of the funding package by the key legislative committees.

The focus on strategic learning moments is similar to Shibley's reminder that information is most likely to be used when it is asked for (Shibley, 1985). A strategic learning moment is a critical time when information or a POE can help resolve a problem or issue that is of considerable importance to the participants. The focus on learning moments can also be used in developing policy documents or targeting POEs toward decisions.

Creating Organizational Memory for Precedents

A key part of organizational memory is simply knowing what the organization has done, but few POE programs have been linked to recording and analysis of precedent. There is a real opportunity to link evaluation to a record of past projects. This record can include simple plans and photos and some analyses of cost, size, and materials. These descriptions can be linked to evaluations.

LESSONS FROM POE PROGRAMS: ENHANCING ORGANIZATIONAL LEARNING

We have suggested that POE has a large potential for lessons learned as well as for project support and analysis. Because of the historic focus of much POE research, the difficulty of finding resources for organizational learning, and sensitivities in exposing problems, relatively few organizations have created effective POE-enabled organizational learning systems that include:

1. monitoring changes in the internal and external business environment,
2. establishing performance goals based on internal and external influences,
3. assessing performance,
4. interpreting and discussing the implications of results,
5. consolidating results into an organizational memory,
6. widely distributing findings and conclusions,
7. creating a culture that allows the organization to take action on the results,
8. taking action based on organizational learning.

However, based on a number of successful examples, we suggest the following strategies for creating the conditions for learning:

- Create opportunities for participation
- Add incentives
- Remove disincentives
- Provide access to information
- Provide upper-level management support

Successful organizations have also used several strategies for creating knowledge:

- Clarify key strategic choices
- Build on existing evaluations
- Focus on strategic moments
- Record precedents

The lessons from these 18 POE programs, which influence billions of dollars of construction, suggest that the solution to creating a lessons-learned program is partly technical, such as using information technology to reduce costs in gathering and distributing information. At its core, however, the problem is organizational—creating a setting in which decisions can be evaluated, discussed, and learned from.

ABOUT THE AUTHORS

Craig Zimring is a professor of architecture and of psychology at the Georgia Institute of Technology. In his teaching, writing, consulting, and research he has developed methods, procedures, and concepts for the evaluation of buildings, including comprehensive studies of building types such as healthcare facilities, jails and prisons, courthouses, and specialized studies of wayfinding, security, stress, and other issues. Dr. Zimring has focused on how social, organizational, and behavioral information can be incorporated into design and decision making at a variety of scales, from a freshman design studio to the $4.5 billion California prison development program, the $6 billion French Universities 2000 program, and the $1 billion annual construction budget of the California Department of General Services. He has worked in the design studio, lectured to facility managers, written in the popular and professional press, served as a consultant and directed research projects for AT&T, U.S. Department of State, U.S. General Services Administration, the Administrative Office of the U.S. Courts, U.S. Department of Transportation, Ministry of Education of France, and many others and served on the boards of several professional organizations including the Environmental Design Research Association and the Justice Facilities Research Program. Dr. Zimring was a distinguished senior visiting fellow at the Centre Scientific et Technique du Batiment, Paris; he has received awards from the American Society of Interior Designers and the National Endowment for the Arts Design Research Recognition Program. He holds a bachelor of science from the University of Michigan, and a master of science and a Ph.D. from the University of Massachusetts at Amherst.

Thierry Rosenheck is a project manager at the U.S. Department of State, Office of Overseas Buildings Operations (OBO), and has been working on embassy rehabilitation projects in New Delhi, Beirut, Tel Aviv, and Jerusalem since 1999. He has developed a serviceability profile of user generic requirements for new chancery office buildings using the *ASTM Standard on Whole Building Functionality and Serviceability* and with the Centre for International Facilities. He has coordinated POE and serviceability input with other ongoing projects at the Office of Overseas Buildings Operations. Prior to working with the Department of State, Mr. Rosenheck was in private practice. He has worked for architectural firms, a construction firm, and taught at the School of Architecture and Urban Planning at Howard University. He holds a bachelor of architecture degree from Howard University, a master's degree in architecture and environment-behavior studies from the University of Wisconsin-Milwaukee, and is a licensed architect in the District of Columbia.

REFERENCES

Argyris, C. (1992a). *On Organizational Learning*. Cambridge, Mass: Blackwell.

Argyris, C. (1992b). Teaching smart people how to learn. In: C. Argyris (Ed.) *On Organizational Learning* (pp. 84-100). Cambridge, Mass: Blackwell Business.

Argyris, C., and Schon, D. (1978). *Organizational Learning*. Reading, Mass: Addison-Wesley.

Baird, G., Gray, J., Isaacs, N., Kernohan, D., & McIndoe, G. (Eds.). (1996). *Building Evaluation Techniques*. New York: McGraw-Hill.

Bechtel, R. (2000). Personal Communication.

Bordass, W., and Leaman, A. (1997). Future buildings and their services: strategic considerations for designers and clients. *Building Research and Information* 25(4): 190-195.

Campbell, D.T. (1999). *Social Experimentation*. Thousand Oaks, California: Sage Publications Inc.

Cohen, R., Bordass, W., and Leaman, A. (1996). *Probe: A Method of Investigation*. Harrogate, United Kingdom: CIBSE/ASHRAE Joint National Conference.

Davis, G., and Szigeti, F. (1996). Serviceability tools and methods (STM): Matching occupant requirements and facilities. In: G. Baird, J. Gray, N. Isaacs, D. Kernohan, G. McIndoe (Eds.) *Building Evaluation Techniques*. New York: McGraw-Hill.

Flagg, G. (1999). Study finds major flaws in San Francisco main library. *American Libraries* 30(9):16.

Friedmann, A., Zimring, C., and Zube, E. (1978). *Environmental Design Evaluation*. New York: Plenum Press.

Gregerson, J. (1997). Fee not-so-simple. *Building Design and Construction* (August): 30-32.

Horgen, T.H., Joroff, M.L., Porter, W.L., and Schon, D.A. (1999). *Excellence by Design: Transforming Workplace and Work Practice*. New York: Wiley.

Huber, G.P. (1991). Organizational learning: The contributing processes and the literature. *Organization Science* 2: 88-115.

Kantrowitz, M., and Farbstein, J. (1996). POE delivers for the Post Office. In G. Baird, J. Gray, N. Isaacs, D. Kernohan, G. McIndoe (Eds.) *Building Evaluation Techniques*. New York: McGraw-Hill.

Kernohan, D., Gray, J., and Daish, J. (1992). *User Participation in Building Design and Management: A Generic Approach to Building Evaluation*. Oxford: Butterworth Architecture.

Leaman, A., Cohen, R. and Jackman, P. (1995). Ventilation of office buildings: Deciding the most appropriate system. *Heating and Air Conditioning* (7/8): 16-18, 20, 22-24, 26-28.

McLaughlin, H. (1997). Post-occupancy evaluations: "They show us what works and what doesn't." *Architectural Record* 14.

Preiser, W. F. E. (1994). Built environment evaluation: Conceptual basis, benefits and uses. *Journal of Architectural and Planning Research* 11(2), 92-107.

Preiser, W.F.E., Rabinowitz, H.Z., and White, E.T. (1988). *Post-Occupancy Evaluation*. New York: Van Nostrand Reinhold.

Preiser, W.F.E., and Schramm, U. (1997). Building performance evaluation. In D.Watson et al. (Eds.) *Time-Saver Standards* (7 ed., pp. 233-238). New York: McGraw-Hill.

Raw, G. (1995). *A Questionnaire for Studies of Sick Building Syndrome*. BRE Report. London: Construction Research Communications.

Raw, G. (2001). Assessing occupant reaction to indoor air quality. In: J. Spengler, J. Samet, and J. McCarthey (Eds.) *Indoor Air Quality Handbook*. New York: McGraw-Hill.

Ripley Architects. (2000). *San Francisco Public Library Post Occupancy Evaluation Final Report*. San Francisco: Ripley Architects.

Schein, E.H. (1995). Learning Consortia: How to Create Parallel Learning Systems for Organization Sets (working paper). Cambridge, Mass: Society for Organizational Learning.

Schneekloth, L.H., and Shibley, R.G. (1995). *Placemaking: The Art and Practice of Building Communities*. New York: Wiley.

Shibley, R. (1982). Building evaluations services. *Progressive Architecture* 63(12): 64-67.

Shibley, R. (1985). Building evaluation in the main stream. *Environment and Behaviour* 1985(1):7-24.

Vischer, J. (1996). *Workspace Strategies: Environment as a Tool for Work*. New York: Chapman and Hall.

Watson, C. (1996). Evolving design for changing values and ways of life. Paper presented at the IAPS14, Stockholm.

Watson, C. (1997). Post occupancy evaluation of buildings and equipment for use in education. *Journal of the Programme On Educational Building* (October).

Zeisel, J. (1975). *Sociology and Architectural Design*. New York: Russell Sage Foundation.

Zimring, C.M., and Reizenstein, J.E. (1981). A primer on post-occupancy evaluation. *Architecture* (AIA Journal) 70(13): 52-59.

Zimring, C.M., and Welch, P. (1988). Learning from 20-20 Hindsight. *Progressive Architecture* (July), 55-62.

6

The Role of Technology for Building Performance Assessments

Audrey Kaplan, Workplace Diagnostics Ltd.

To date, building performance assessments have been limited, for the most part, to large organizations and institutions. The substantial investment of time and money needed to mount such assessments using traditional methods is a major hurdle for most midsize and small organizations or for niche groups. Relative to the number of facilities and organizations that could benefit from such assessments, too few are done to make them an effective tool for aligning occupant or stakeholder expectations with the finished product. Advances in electronic communications and easy-to-use Web browsers now make it attractive to conduct these assessments via the Internet or an intranet. "Cybersurveys" or "e-surveys" (polls or assessments administered electronically) represent the breakthrough in social science research that will make this work cheaper and more effective to complete. Moreover, the innovative medium will likely inspire the invention of wholly new work and objectives that could not be done using existing assessment methods (Bainbridge, 1999).

Regardless of the medium, all assessment work begins with good research design, clearly stated objectives, solid planning, and preparation to conduct a survey, complete the analysis, and produce reports. This chapter addresses issues related to implementing cybersurveys and assumes that sound principles of building performance assessment are in place. Information technology has unique features to gather feedback about building performance, which in turn can be used to improve facility management and acquisition.

As the set of Internet users begins to reflect the population in general, or a specific group under study, cybersurveys may become the predominant method of administering building assessments. If widely used, the assessment process can be an effective tool for continually improving the value of each facility, regardless of its size or unique occupancy.

INTRODUCTION

Cybersurveys complement existing survey methods of assessment, such as paper, mail, phone, in-person interview, site visits, expert observations, photography, and instrument measurements.

Instruments to assess buildings continue to improve steadily. Prompted by the energy crisis of the 1970s, instruments were developed to measure and/or diagnose the effectiveness of building systems (e.g., walls; windows; roofs; heating, ventilation, and air conditioning; lighting). Generic methodologies were conceived to diagnose total building systems, and tools were designed to assess the performance of specific building aspects. See, for example, *A Generic Methodology for Thermographic Diagnosis of Building Enclosures* (Mill and Kaplan, 1982) for a methodology and tools to determine heat loss and uncontrolled air flow across building enclosures. The objective of methodologies and tools developed at that time was to better understand the cause of deficiencies—be they rooted in design or construction—and then to prescribe corrective actions and processes to prevent future occurrences.

In the 1980s, the heightened concern for the quality of interior environments drove the development of new tools to better represent the total environmental setting that occupants experience. Many traditional lab instruments were modified (or put on carts) and brought into occupied buildings. The focus was to record environmental factors that influence human comfort and performance (e.g., light, sound, temperature, relative humidity, air quality, spatial configuration, aesthetics). These instruments were often cumbersome or too delicate for such robust applications, so manufacturers and researchers alike redesigned them to better match the field conditions and the nature of the data being collected (see, for example, Schiller et al., 1988, 1989). Some of the features modified to make reliable field instruments included narrowing the instruments' range, adjusting the sample duration and frequency, placing sensors to record levels in occupied zones (as opposed to air supply ducts, etc.), and immediate statistical analysis of select field data for preliminary interpretation to guide the balance of the data collection. Since then, there have been steady, incremental improvements to the set of field instruments, benefiting from miniaturization and faster processing as the whole computer industry developed.

Academic centers continue to advance the use of instruments to assess building performance. See, for example, the Center for Environmental Design Research (also known as the Center for the Built Environment) at the University of California, Berkeley, www.cbe.berkeley.edu, and Cornell University, Ithaca, New York, ergo.human.cornell.edu. Dependent upon funding to support building performance programs, these and other institutes lend instruments and materials to architecture, engineering, and facility management schools. These lending programs are intended to encourage the next generation of built environment professionals to build and manage environmentally responsible and energy-efficient structures. For example, the Vital Signs program at the University of California, Berkeley (1996), assembled tools to measure a range of building performance attributes. These were packaged into kits that educators could borrow for their classes. Support funding has since expired, however other programs carry on toward the same objectives. See, for example, the occupant survey project <www.cbe.berkeley.edu/cedr/cbe>.

Improvements continue in the measurement of the physical built environment, but there are no major breakthroughs or fundamental changes in how or what is measured. Similarly, important advances in opinion research, such as telephone interviews, did not fundamentally change the way data were collected and analyzed or questionnaires were designed (Taylor, 2000). However, the visual and responsive characteristics of cybersurveys offer new and significant opportunities in the way building assessments are conceived and completed. The author sees cybersurveys as the next technology-based advance in building performance assessments. The remainder of this chapter discusses how this technology is used to assess building performance.

CYBERSURVEYS

The unique features of computer technology (e.g., Web, e-mail, electronic communications) improve upon and add efficiency to the polling process. The traditional methods of surveying people—telephone, mail, or in person—are expensive and tend to be used now only by large organizations. The negligible cost to distribute surveys over the Internet or intranet and the potential to reach far and wide are changing the survey industry. Now, small and medium-size businesses use cybersurveys as a means to gather valuable information from customers, potential customers, employees, or the general public. Readily available survey development products help the nonspecialist to build a Web survey questionnaire, publish it on the Web, collect the data, and then analyze and report the results (see Appendix E). These tools address how to conduct the research and the mechanics of data collection. They do not, and cannot, replace the management decision to conduct the inquiry, how to interpret the numbers, or how to get actionable results from the effort.

Cybersurveys are distributed as e-mail to a specific address (either embedded in the message or as an attachment) or openly on the Web. There are more control and layout options with Web-based surveys than with e-mail. With HTML (hypertext markup language), attractive and inviting forms can be created. Some of the survey software packages have automated routines to design the survey's physical appearance.

The cost of initial data collection is high—whether for personnel to conduct interviews and code data or for setting up a computer-based survey to automatically solicit and compile data. However, the biggest issues currently facing cybersurveys are the low response rate (compared to traditional methods and relative to the potential based on number of visits to a site) and ensuring a proper sample. These issues are discussed in the next two sections, followed by lessons learned from Web-based surveys.

RESPONSE RATE

It is widely agreed that the more attempts made to contact respondents, the greater are the chances of their returning a survey. Similarly, a greater number of response formats increases the likelihood of more responses. Multiple contacts offering alternate ways of responding with appropriate personalization, singly and in combination, increase response rate. For example, Schaefer and Dillman (1998) reported from published studies the average response rate for e-mail surveys. With a single contact, the response rate was 28.5 percent, with two contacts it was 41 percent, and with three or more contacts the response rate was 57 percent. Cybersurveys offer another contact vehicle and method of response (e.g., interactive display, print and return) to the options already available. In and of itself, use of this vehicle contributes to higher response rates.

A consistent finding is that cybersurveys are returned more quickly than mail responses. In comparing several studies that used both e-surveys and paper (mail and fax), response speeds ranged from 5 to 10 days for electronic distribution versus 10-15 days for mail (Sheehan and McMillan, 1999).

To date, response rates from e-surveys have been lower than for postal mail. E-mail rates range from 6 to 75 percent. However the electronic and mail surveys used for this comparison (Sheehan and McMillan, 1999) had small sample sizes (less than 200) and varied widely in survey topic and participants' characteristics, so the spread is not surprising. E-surveys often achieve a 20 percent response, which is half of the rate usually obtained by mail or phone surveys. Some e-surveys have had less than 1 percent return of the possible responses, based on the number of hits to a Web site (Basi, 1999). There were some exceptions to this trend in the early 1990s when e-surveys were distributed to workers in electronic-related businesses (e.g., telephone and high-tech sectors). At that time, there was still a high-tech allure and/or novelty to receiving e-mail, so response rates were acceptable (greater than 60 percent). This highlights how quickly the field is changing. Lessons learned from early cybersurveys may not necessarily apply in a current context. See Appendix E for a discussion of the changing context of on-line communications.

Other reasons for low participation in on-line surveys may involve a reluctance to share one's views in a nontraditional environment. However, for those who are comfortable with this arrangement, replies to open-ended questions are richer, longer, and more revealing than those from other methods (Taylor, 2000). Cybersurveys are nontangible. Paper surveys may sit on someone's desk and serve to remind him or her to complete them. Individuals intending to complete a cybersurvey may bookmark the survey's URL (uniform resource locator) for later use but then forget to retrieve it.

SAMPLING

The Web is worldwide. People anywhere on the planet can answer a survey posted on the Web even if they are not part of the intended audience (unless controls are used, such as a unique URL, a password, or setting the meta tags so that the page is not picked up by search engines). To date, cybersurveys are completed by accidental volunteers (they stumbled upon the survey) and self-selected individuals who choose to participate.

Because there is no central registry of Web users, one does not know who is on-line, so one cannot reliably reach a target population (see Appendix E for details of who is on-line and where they are geographically). Moreover, there is no control on how electronic messages are cascaded or linked to other sites. A message can be directed to select individuals or groups who, for their own reasons, send it along to others. This behavior makes it difficult to ensure that the survey reaches a random or representative sample of respondents. Despite these sampling difficulties, careful use of nonprobability sampling can produce results that represent a specific subset of the population (Babbie, 1990).

LESSONS LEARNED

The following recommendations for Web surveys are adapted from published papers, Web sites, and the author's experience (Bradley, 1999; Kaye and Johnson, 1999; Sheehan and McMillan, 1999; Perseus, 2000). Recommendations are organized under the topics of Web survey design considerations, sampling, publicity, and data collection and responses.

Web Survey Design Considerations

1. Keep the survey as short as possible for quick completion and minimum scrolling.

TABLE 6-1 Internet Users, by Region, Who Prefer to Use the Web in English Rather than Their Native Tongue.

Region	% Who Prefer English
North America	74
Africa, Middle East	72
Eastern Europe	67
Asia	62
Western Europe	51
Latin America	48

Source: USA Today (2001).

2. Use simple designs with only necessary graphics to save on download time.
3. Use drop-down boxes to save space, reduce clutter, and avoid repeating responses.
4. State instructions clearly.
5. Give a cutoff date for responses.
6. Personalize the survey (e.g., use recipient's e-mail address in a pre-notification letter; identify the survey's sponsor, which can also add credibility).
7. Assure respondents that their privacy will be protected—both their e-mail address and the safety of their computer system (virus free).
8. Conduct pre-tests to measure time and ease for completing the survey. Electronic pre-testing is easier and cheaper than traditional methods.
9. Try to complete the survey using different browsers to uncover browser-based design flaws.
10. English-language cyber-surveys can easily overcome geographic barriers (Swoboda et al., 1997). See Table 6-1 for the percentage of Internet users who prefer to use the Web in English rather than their native language.

Sampling

1. To generalize the results to a wider population, define samples as subsets of Web users based on some specific characteristic.
2. Solicit responses from the target population. This can be done by linking the survey from select sites, by posting announcements on discussion-type sites that the target population is likely to use, or by selecting e-mail addresses posted on key Usenet newsgroups, listserves, and chat forums. Invited participants should be given an identification number and password, that are both required to access the Web-based survey.
3. Clearly state the intended audience in the introduction to the survey so, hopefully, only those to whom it applies will respond.

Publicity

1. Systematically publicize the survey daily through various means. Create awareness of the survey at a wide variety of Internet or intranet outlets to reduce bias. "Pop-up" surveys may attract more responses than simple banner invitations that are fixed to a page. Advertising on select sites might increase the number of completions but does not continually invite (spam) a few discussion groups while ignoring others. See Table 6-2 for methods to invite people to a survey.
2. Pre-notify respondents who would find the survey relevant (e.g., those to whom an important issue is current or timely).
3. Do not go overboard on publicity, just announce the survey.
4. List the survey with as many of the major search engines as possible. Services, such as "Submit It," send listings to many search engines with just one entry. Then use different search strategies and terms to locate the survey and uncover glitches or errors.

TABLE 6-2 Methods to Attract Respondents to a Cybersurvey.

Announcements	Browsing intercept software	Customer records
E-mail directories	Harvested addresses	Hypertext links
Interest group members	Invitations (banners, letters, etc.)	Pop-up surveys
Printed directories	Registration forms	Snowballing
Staff records	Subscribers	Web site directories

Source: Bradley (1999).

5. Check that the survey links remain posted and that the URL is visible on the page.

6. Write the complete URL in announcements or messages because it can be an easy, "clickable" link in some e-mail transmissions.

7. Ask respondents how they found out about the survey to gauge which sites and discussion outlets were most effective in reaching the target audience.

8. If appropriate and/or feasible, offer incentives for completing the survey (ranging from the survey results to lottery tickets, money, or discounts for a third-party product). Check the laws governing incentives because these vary by jurisdiction.

Data Collection and Responses

1. Design the survey so that it is submitted with a simple click of the mouse or keystroke.

2. Upon receiving the survey, set up for an automatic thank-you reply to the sender saying that the survey was successfully transmitted.

3. To check for duplicate responses, ask for respondents' e-mail address and/or track senders' Internet or intranet protocol address.

4. For ease of importing e-mailed data into a statistical software program, design the return forms so that each question is listed on one line followed by its response and a corresponding numerical value.

5. People may respond to electronically viewed scales differently than to scales that are spoken or on paper. Taylor (2000) observed that fewer people picked the extremes on e-scales than when they heard the questions and response options.

CONCLUSIONS AND DISCUSSION

The Internet or intranet and World Wide Web enable traditional building performance assessments to be accomplished more cheaply and effectively. The electronic medium may well become the primary survey vehicle owing to its convenience, low-cost of distribution and return, ability to verify (check for errors) and receive data—including rich text replies—in electronic format, and the fact that it is an easy way to give respondents feedback. All this is done incredibly fast with an attractive, visual presentation.

The opportunities presented by the medium itself will likely lead to the invention of new products,

methods, and presentations for building performance assessments. For example, new lines of research might be conceived that integrate occupants' perception of interior environments with data from the building control system and/or weather conditions or that track the use of facilities as recorded by smart cards and/or the spatial coordinates of workspaces as recorded by computer-aided drafting systems.

Select opinion leaders and/or professionals could be drawn to an Internet-based "brain trust" that focuses on special questions related to the built environment. The Internet or intranet and Web are excellent media for establishing virtual libraries and databases that many can access and use. Establishing sites with existing data allows other researchers to produce interpretations, conclusions, or knowledge that is in some way different from that produced in the original inquiry (i.e., secondary analysis). This work would economically and efficiently add to the experience and knowledge of building performance assessments.

With cybersurveys, questions can be asked of anyone, worldwide, who is on-line. It is feasible to poll a wide population about issues that impact a specific building. The sample is not limited to those who work in, visit, or are somehow involved with a building. Tapping into these opinions becomes important when there is a highly visible construction, a new and important public structure, or a controversy connected with a facility. Public opinion might provide valuable insight to how a building, its occupants, or a particular issue is viewed. For example, an incidence of sick-building syndrome or an environmental disaster might be officially resolved to those managing a facility, but the local community could remain suspicious and unsupportive of the organization. Using wide-reaching cybersurveys to tap into that sentiment gives insight that might influence decisions regarding the building's publicity, image, operational management, and the communication to users and other interested parties of what was done.

There is great value in being able to reach a population that is not yet at a facility. For example, high school students are actively on-line and, through cybersurveys, can be asked their expectations for their university experience—how they will live, learn, study, and make friends. It would be insightful to determine that they have, for example, no desire to study in traditional libraries or in their dormitory rooms. They may expect "cybercafés," collaborative work settings, or learning environments that ease their transition into work environments where they will earn a livelihood. Through select demographic questions, survey researchers can characterize the future university population and suggest built environment features compatible with their expectations. Universities may use these as features to attract top students and then to elicit the best work from enrolled students.

Much of the young, skilled work force in the technology sector comes from outside North America and from different cultures. Each has its own expectation of work and the workplace. Here again, using cybersurveys to understand their expectations of the workplace and to educate them about North American workplaces can speed their integration into the work force. There is a strong business case to support this use of cybersurveys since they readily translate into business objectives of worker productivity and reduced time to market for products.

Occupant feedback is difficult to get, or unreliable, when people are stressed, such as when they need hospitals, senior residences, and funeral parlors. However, after the crisis has passed and their emotions are stable, they could offer insight to aspects of the physical setting that helped them at the time or that would have been of benefit. Cybersurveys can reach these people at a time when they can offer measured opinions on the built environment.

The rewards of cybersurveys are so rich that their potential will surely be used and the methodological and sampling issues of today will be resolved. It is hard to imagine future building performance assessments without extensive use of the Internet or intranet and the Web. However, traditional methods will continue to be used, perhaps becoming a smaller part of the overall work. Printing did not fully replace handwriting, radio did not take the place of newspapers, and television did not supplant movies or radio. Cybersurveys enable many things that could not be done or afforded using traditional methods. The addition of this medium to the tools for building performance assessment greatly enhances the accessibility and value of such data and of the evaluation process itself.

ABOUT THE AUTHOR

Audrey Kaplan is president of Workplace Diagnostics, Ltd., an Ottawa-based consulting company that specializes in the evaluation and design of workspace. Ms. Kaplan has been actively involved in building performance as a research scientist and consultant for 20 years and has published widely in the field. Her research activities range from the evaluation of comfort and environmental quality in offices, to the assessment of total building performance, to the design of workstations with personal environmental controls. In addition to consulting, Ms. Kaplan is a regular speaker at professional conferences and continuing education seminars, and is a certified instructor for professional training courses with the International Facility Management Association (IFMA). She was an adjunct professor at the University of Manitoba and served on IFMA's international board as a director from 1998-2000, as chair of IFMA's Canadian Foundation; and as a trustee of IFMA's foundation. Ms. Kaplan received IFMA's 1996 Distinguished Author Award for her co-authored book *Total Workplace Performance: Rethinking the Office Environment*. She holds a bachelor of architecture from Carleton University and a master of science in architecture from Carnegie Mellon University.

REFERENCES

Babbie, E. (1990). *Survey Research Methods*. Belmont, California: Wadsworth.

Bainbridge, W. (1999). Cyberspace: Sociology's natural domain. *Contemporary Sociology* 28(6):664-667.

Basi, R. (1999). WWW response rates to socio-demographic items. *Journal of the Market Research Society* 41(4):397-401.

Bradley, N. (1999). Sampling for Internet surveys: An examination of respondent selection for Internet research. *Journal of the Market Research Society* 41(4):387-395.

Center for Environmental Design Research (known as the Center for the Built Environment) (1996). *Vital Signs*. Berkeley: University of California (www.cbe.berkeley.edu).

Kaye, B., and Johnson, T. (1999). Research methodology: Taming the cyber frontier. *Social Science Computer Review* 17(3):323-337.

Mill, P., and Kaplan, A. (1982). *A Generic Methodology for Thermographic Diagnosis of Building Enclosures*. Public Works Canada. Report Series No. 30.

Perseus. (2000). Survey 101—A complete guide to a successful survey (www.perseus_101b.htm).

Pike, P. (2001). Technology rant: I'm tired of feeling incompetent! *PikeNet* February 11 (www.pikenet.com).

Schaefer, D., and Dillman, D. (1998). Development of a standard e-mail methodology. *Public Opinion Quarterly* 62:378-397.

Schiller, G., et al. (1988). *Thermal Environments and Comfort in Office Buildings*. Berkeley: Center for Environmental Design Research, CEDR-02-89

Schiller, G., et al. (1989). Thermal comfort in office buildings. *ASHRAE Journal* October 26-32.

Sheehan, K., and McMillan, S. (1999). Response variation in e-mail surveys: An exploration. *Journal of Advertising Research* 39(4):45-54.

Swoboda, W., Muhlberger, N., Weitkunat, R., and Schneeweib, S. (1997). Internet surveys by direct mailing. *Social Science Computer Review* 15(3):242-255.

Taylor, H. 2000. Does Internet research work? *International Journal of Market Research* 42(1):51-63.

USA Today (2001). Searching the Web in native language. February 27, p. 7B.

Appendixes

Appendix A

Functionality and Serviceability Standards: Tools for Stating Functional Requirements and for Evaluating Facilities

Françoise Szigeti and Gerald Davis, International Centre for Facilities

INTRODUCTION: THE FUNCTIONALITY AND SERVICEABILITY TOOLS HAVE STRONG FOUNDATIONS[1]

The functionality and serviceability tools are founded in part on "the performance concept in building," which has roots before World War II in Canada, the United States, and overseas. In the United States in the 1950s and 1960s, the Public Buildings Service (PBS) of the General Services Administration (GSA) funded the National Institute of Standards and Technology (NIST, then the National Bureau of Standards) to develop a performance approach for the procurement of government offices, resulting in the so-called Peach Book publication (NBS, 1971). Starting in the early 1980s, the performance concept was applied to facilities for office work and other functions by the American Society for Testing and Materials (ASTM) Subcommittee E06.25 on Whole Buildings and Facilities. Worldwide, in 1970, the International Council for Building Research Studies and Documentation (commonly known as CIB) set up Working Commission W060 on the Performance Concept in Building. In 1982, the coordinator for that commission defined the concept in those terms: "The performance approach is, first and foremost, the practice of thinking and working in terms on ends rather than means. It is concerned with what a building is required to do, and not with

prescribing how it is to be constructed" (Gibson, 1982). In 1998, the CIB launched a proactive program for the period 1998-2001 focused on two themes: the performance-based building approach, and its impact on standards, codes and regulations, and sustainable construction and development.[2]

By 1985, the importance of distinguishing between *performance* and *serviceability* had been recognized, and standard definitions for *facility* and *facility serviceability* were developed. *Facility performance* is defined by ASTM as the "behaviour in service of a facility for a specified use," while *facility serviceability* is the "capability of a facility to perform the function(s) for which it is designed, used, or required to be used." Both definitions are from ASTM Standard E1480. Serviceability is more suited than performance to responding to the stated requirements for a facility, because the focus of performance is only on a single specified use or condition, at a given point in time, whereas serviceability deals with the capability of a facility to deliver a range of performance over time. In the International Organization for Standardization (ISO), related work has been carried out within ISO/Technical Committee 59/Sub-Committee 3 on Functional/User Requirements and Performance in Building Construction.

The term *programme*, meaning a statement of requirements for what should be built, was in common usage in the mid-nineteenth century by architectural students at the Ecole des Beaux Arts in Paris and, thereafter, in American universities as they adopted the French system. In North America, the architect's basic

---

[1]For further information and details, see Szigeti and Davis, (1997) in Amiel, M. S., and Vischer, J. C., *Space Design and Management for Place Making* Proceedings of the 28th Annual Conference of the Environmental Design Resarch Association, The Environmental Design Research Association (EDRA), Edmond, Okla., 1997.

[2]CIB Pro-Active Program, see CIB Web site for further details at <www.cibworld.nl>.

services included architectural programming (i.e., "confirming the requirements of the project to the owner"), but excluded setting functional requirements, which was the owner's responsibility. In Britain and parts of Canada, the term "briefing" includes programming. By the mid-twentieth century, some clients for large or complex projects paid extra to have their architects or management consultants prepare a functional program for their projects.[3] The functionality and serviceability tools were created to make it easier, faster, and cheaper to create such functional programs in a consistent and comprehensive manner, to link requirements to results, and to evaluate performance against requirements.

There is now a worldwide trend toward the use of a "performance-based approach" to the procurement, delivery, and evaluation of facilities. This approach is useful because it focuses on the results, rather than on the specification of the means of production and delivery. It reduces trade barriers and promotes innovation or at least removes many impediments to innovation. For such an approach to be successful however, there is a need for more attention to be paid to the definition and description of the purposes (demand-results), short term and long term, and for more robust ways of verifying that the results have indeed been obtained. This is why there is a mounting interest in building performance evaluations and other types of assessments. Customer satisfaction surveys, post-occupancy evaluations (POEs), lease audits, and building condition reports are becoming more common.

This appendix has four sections: (1) context, (2) measuring the quality of performance of facilities using the ASTM standards, (3) examples, and (4) final comments.

CONTEXT: PROGRAMMING AND EVALUATION AS PART OF A CONTINUUM

Feed-Forward—The Programming-Evaluation Loop

Not only do most large organizations lack a comprehensive facilities database, they also fail to develop an

institutional memory of lessons learned. They are too often dependent on what best practices have been recognized and remembered by individual real estate and facility staff members and passed on informally to their subordinates and successors. Most often, such accumulated knowledge disappears with the individuals responsible. Instead, as each facility project is acquired, whether it is new construction, remodel or refit, or leased or owned, both the facility and the processes involved should be evaluated. Each phase of each project should be considered a potential source of lessons, including planning, management, programming, design, construction, commissioning, occupancy, operation, and maintenance, even decommissioning. Figure A-1 shows such an ongoing cycle of feed-forward from project to project.

To be effective, such evaluations, or programs of lessons learned, need a way to organize the information and to relate and compare it to what the client requires now and in the future. Since 1965 when TEAG—The Environmental Analysis Group/GEMH – Groupe Pour l'Etude du Milieu Humain—was launched, a programming assignment normally starts with an evaluation of the current facilities used by the client or similar surrogate facilities if need be. These evaluations give invaluable information and serve as a context for the programming process. The work then proceeds with interviews of senior managers about current problems and future expectations, and group interviews with occupants at several levels of the organization. Questions are asked about what works, not just what does not work. It is important to know what should be carried over from the current situation.

Over the years, interview guides and recording documents have been developed for such evaluations. This work and experience provided the foundation for the functionality and serviceability tools. Thus, evaluations feed into functional programs, which become the basis for the next evaluations.

Defining Requirements

The functional program should focus on aspects of the project requirements that are important for the enterprise, in order to direct the best allocation of resources within the given cost envelope. The objective is to get best value for the users and owners. A knowledgeable client will prepare, in-house or with the help of consultants, a statement of requirements (SOR), including indicators of capability of the solution that

---

[3]The first stand-alone, general practice in building programming, not part of an architectural or management consulting practice, was TEAG—The Environmental Analysis Group, founded in 1965 by Gerald Davis. The International Centre for Facilities was founded in 1986 to focus on research and development activities related to facilities and on standard development activities.

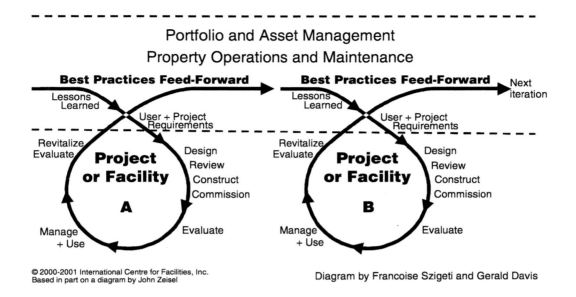

Portfolio and Asset Management
Property Operations and Maintenance

Best Practices Feed-Forward

Lessons Learned

User + Project Requirements

Revitalize Evaluate

Project or Facility A

Design
Review
Construct
Commission

Manage + Use

Evaluate

Best Practices Feed-Forward

Next iteration

Lessons Learned

User + Project Requirements

Revitalize Evaluate

Project or Facility B

Design
Review
Construct
Commission

Manage + Use

Evaluate

© 2000-2001 International Centre for Facilities, Inc.
Based in part on a diagram by John Zeisel

Diagram by Francoise Szigeti and Gerald Davis

FIGURE A-1 Feed-forward.

are easy to audit and are as unambiguous as practicable. This is an essential step of the planning phase for a project.

Portfolio management provides the link between business demands and real estate strategy. At the portfolio level, requirements are usually rolled up and related to the demands of the strategic real estate plan in support of the business plan for the enterprise (Teicholz, 2001).

Requirements for facilities needed by an enterprise will normally be included in a portfolio management strategy. An asset management plan for a facility would include the specific requirements for that facility. A statement of requirements, in one form or another, more or less adequate, is part of the contractual documentation for each specific procurement.

Statements of requirements serve as the starting point for providers of material, products, facilities, services, and so forth. As experienced readers of this report likely know, if there is ever litigation or other liability issue, then the statement of requirements is the first document that the parties will turn to. It is the reference point for any review process during tendering, design, production, and delivery, as well as for later evaluation(s), no matter which methodology is used. It

is particularly necessary for performance-based and design-build procurement and for any project developed using an integrated project team approach. In the more traditional approach, the contractual documents normally include very precise specifications ("specs"). In the experience of expert witnesses, the root of many court cases and misunderstandings can be traced back squarely to badly worded, imprecise, incomplete statements of requirements that do not include any agreed means of verifying whether the product or service delivered is in fact meeting the stated requirements.

In a performance-based approach and for design-build and similar procurements, the focus is on the expected performance, or on a range of performances, of the end product. Therefore, the heart of these nontraditional approaches is defining those expected results and the requirements of the customer or user in an objective, comprehensive, consistent, and verifiable manner. In any dispute, it is necessary to be able to go back to the contract and have a clear definition of what was agreed between the parties. If the "legal" name of the game is a "warranty of fitness for purpose," then the purpose has to be clearly spelled out, as well as the ways to verify that "fitness." This point is developed further later in this appendix.

Life Cycle of Facilities, Shared Data, and Relationship to the Real Estate Processes of the Enterprise

For each facility, the information included in the asset management plan, plus the more detailed programming data and the financial data, are the foundation for the cumulative knowledge base of shared data and support data about the facilities diagrammed at the center of Figure A-2. Throughout the life cycle of a facility, many people, such as portfolio and facility managers, users, operations and maintenance staff, financial managers, and others, should be able to contribute to and access this pool of data, information, and knowledge.

Today, these kinds of data and information are still mostly contained in "silos," with many disconnects between the different phases of the life cycle of a facility. Too often the data are captured again and again, stored in incompatible formats and difficult to correlate and keep accurate. The use of computerized databases and the move to Web-based software applications and projects are steps toward the creation of a shared information base for the management of real estate assets. Once such shared databases exist, the value of evaluations and benchmarking exercises will increase because the information will be easier to retrieve when needed. The shared data and knowledge base will also make it easier to " close the loop" and relate the facilities delivered to the demands of the enterprise.

In discussions at the Facilities Information Council of the National Institute of Building Sciences, such recreation of data over and over again has been identified as a major cause of wasted dollars and the source of potential savings. More important will be the reductions in misunderstandings, the increased ability to pin-

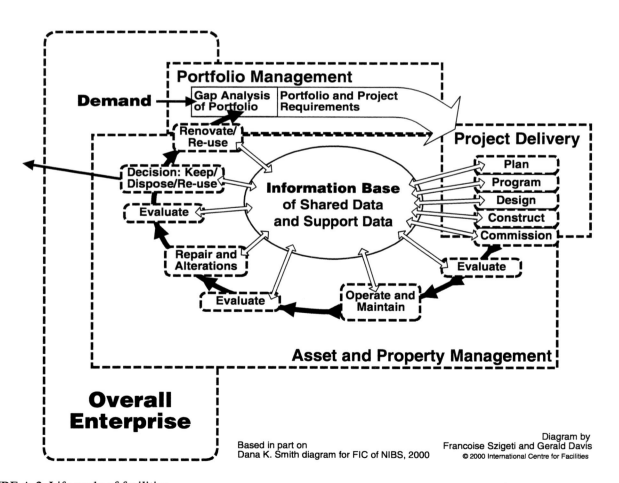

Based in part on
Dana K. Smith diagram for FIC of NIBS, 2000

Diagram by
Francoise Szigeti and Gerald Davis
© 2000 International Centre for Facilities

FIGURE A-2 Life cycle of facilities.

point weak links in the information transfer chain, and the improvements in the products and services because the lessons learned will not be lost.

For evaluations to yield their full potential as part of the life-cycle loop, the information that is fed forward from such activities needs to be captured and presented in comparable formats. Accepted terminology, standard definitions, and normalized documentation will make such comparisons much easier. The links between evaluations and stated requirements should be explicit and easy to trace.

Figure A-2 diagrams the life cycle of a facility, including the particular points in the cycle when most evaluations occur. It shows in greater detail how one of the feed-forward loops unfolds. During project delivery, of course, there are or should be evaluation loops that cannot be shown in the overall diagram.

Evaluations: From Strategic to In-Depth

Evaluations can happen at any time and can be triggered by many situations. They range in their approach from the perceptual to the performance-based to the specific and technical. At the moment, there is not yet any consensus as to what kinds of evaluations should be done and when or how they should be done. Even the terminology is in great flux. Many terms are used, such as analysis, assessment, audit, evaluation, investigation, rating, review, scan, and so forth, usually together with some qualifier, such as building condition report, building performance evaluation, facility assessment, post-construction evaluation, post-occupancy evaluation, serviceability rating, etc.

The range of tools, methods, and approaches to evaluations is quite wide as well as deep. In some situations, it is appropriate to take a broad strategic view and to use tools that can give answers quickly and with the minimum of effort. At the other extreme, there are situations that call for in-depth, specific, narrowly focused, very technical engineering audits that can take weeks and require sophisticated instrumentation. Figure A-3 shows the relationship between these different levels of precision.

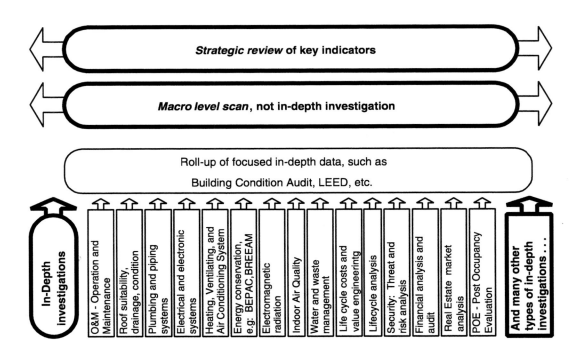

FIGURE A-3 Strategic to in-depth evaluations. Source: Francoise Szigeti and Gerald Davis, © 1999, 2000 International Centre for Facilities.

MEASURING AND MANAGING THE *QUALITY* OF PERFORMANCE OF FACILITIES USING THE ASTM STANDARDS ON WHOLE BUILDING FUNCTIONALITY AND SERVICEABILITY

Assessing Customer Perception and Quality of Performance

Assessing customer perception and satisfaction, or evaluating the quality of the performance delivered by a facility in support of customer requirements are two complementary, but *not* identical, types of assessments. In a recent issue of *Consumer Reports*, there is a series of items dealing with the ratings of health maintenance organizations. In one of the articles, the question of the quality of the ratings is posed and an important point is made. "Satisfaction measures are important. But, don't confuse them with measures of medical quality..." (Consumer Reports, 2000). The key point is that measuring customer satisfaction is important and necessary, but not sufficient.

There is a need for measuring the actual *quality* and performance of the services and products delivered, whether it be medical care or the facilities and services they provide in support of the occupants and the enterprise.

Defining Quality

Quality is described in ISO 9000 as the "totality of features and characteristics of a product or service that bear on its ability to satisfy stated and implied needs" (ISO, 2000). Quality is also defined as "fitness for purpose at a given cost." The difference between Tiffany quality and Wal-Mart quality does not need to be explained. Both provide quality and value for money. Both are appropriate, depending on what the customer is looking for, for what purpose, and at what price.

Quality therefore is not absolute. It is the most appropriate result that can be obtained for the price one is willing to pay. Again, in order to be able to evaluate and compare different results or offerings, and verify whether the requirements have been satisfied, these must be stated as clearly as possible.

Measuring and Managing the Quality of Performance

Many enterprises, public and private, review the project file during commissioning, or later, and note whether the project was completed within budget and

on schedule. Some do assess how well each new or remodeled facility meets the need of the business users who occupy it. Essential knowledge can be captured as part of a formal institutional memory of what works well, what works best, and what should not be repeated. There are an array of different methods and tools that can be used to capture this information. A number of those tools have been catalogued by a group of researchers and practitioners based at the University of Victoria at Wellington, New Zealand (Baird et al., 1996).

A quality management (or assurance) program needs to measure and track performance against "stated requirements." Those who provide a product or service (e.g., a facility and its operations and management), should ascertain the explicit and implicit requirements of the customers (occupants), decide to what level those needs should be met, meet that level consistently, and be able to show that they are in fact meeting those requirements within the cost envelope.

Such programs, therefore, need to start with an appropriate process for preparing statements of requirements. These should include the ability to determine and assess features and characteristics of the product or service considered; to relate them directly to customers' needs, expectations, and requirements; and to document it all in a systematic, comprehensive, and orderly manner. Such documentation should include the means to monitor compliance during all phases of the life cycle of the facility. When dealing with facilities, information should also be included about how the enterprise is organized and its business strategy, and about expectations related to quantity, constraints, environmental and other impacts, time, costs, and so forth. All these elements have to be taken into consideration when conducting an overall evaluation, in particular at the time of commissioning or shortly thereafter.

Using the ASTM Standards on Whole Building Functionality and Serviceability to Measure Quality

The information provided by most POEs and by customer satisfaction surveys is primarily about occupant perception and satisfaction, which is often necessary but rarely sufficient. It is seldom specific enough to be acted upon directly. Similarly, in-depth and specific technical evaluations usually do not address topics directly related to the functional requirements of the users or cannot be matched to those requirements.

Based on some 30 years of experience with both programming and evaluation, over the period 1987-1993, the International Centre for Facilities (ICF) team created a set of scales that have now become the ASTM standards on whole building functionality and serviceability, recognized as American National Standards (ANSI). More importantly, these standards are based on a methodology for creating such scales that is currently being balloted at the international level under the authority of ISO TC 59/SC3 on Functional/User Requirements and Performance in Building Construction (ASTM, 2000).

These standards do provide information that can be acted upon. They measure the quality of services delivered by the facility in support of the occupants as individuals and as groups. The results from serviceability ratings complement POEs and can be cross-referenced to customer satisfaction surveys (see examples in "Final Comments"). These standards currently provide explicit, objective, consistent methods and tools and include the means to monitor and verify compliance with respect to office facilities. The usefulness of such structured information goes beyond a single project. It can also be used for lessons-learned programs and for benchmarking. The methodology could be applied equally well to create a set of tools for measuring the quality of performance of any capital asset, including all types of constructed assets, whether public infrastructure such as bridges, roads, and utilities, or buildings.

This new generation of tools gives real estate professionals the means to evaluate the "fit" between facilities and the users they serve. These tools use indicators of capability to assess how well a proposed design, or an occupied facility, meets the functional requirements specified by the business units and facility occupants. Even a small business, with only a few dozen staff, needs to capture and conveniently access the key facts about its workplaces, how they are used, and lessons to apply "next time."

Functionality and Serviceability: Matching User Requirements (Demand) and Their Facilities (Supply)

Evaluations are most useful when they provide the means to compare results to expectations. Figure A-4 shows the use of bar-chart profiles to match functional requirements and serviceability ratings using the ASTM standard scales.

ASTM Standard Scales

The ASTM standard scales provide a broad-brush, macro level method, appropriate for strategic, overall decision-making. The scales deal with both *demand* (occupant requirements) and *supply* (serviceability of buildings) (MacGregor and Then, 1999). They can be used at any time, not just at the start of a project. In particular, they can be used as part of portfolio management to provide a unit of information for the asset management plan, on the one hand, and for the roll-up of requirements of the business unit, on the other.

The ASTM standard scales include two matched, multiple-choice questionnaires and levels. One questionnaire is used for setting workplace requirements for functionality and quality. It describes customer needs—*demand*—in everyday language, as the core of front-end planning. The other, matching questionnaire is used for assessing the capability of a building to meet those levels of need, which is its serviceability. It rates facilities—*supply*—in performance language as a first step toward an outline performance specification.

Both cover more than 100 topics and 340 building features, each with levels of service calibrated from 0 to 9 (less to more). These standard scales are particularly suitable as part of the front end for a design-build project, to compare several facilities on offer to buy or lease. The scales can also be used to compare the relative requirements of different groups.

This set of tools was designed to bridge between "functional programs" written in user language on the one side and "outline specifications and evaluations" written in technical performance language on the other. Although it is a standardized approach, it can easily be adapted and tailored to reflect the particular needs of a specific organization.

For organizations with many facilities that house similar types of functions, the functionality and serviceability scales capture a systematic and consistent record of the institutional memory of the organization. Their use speeds up the functional programming process and provides comprehensive, systematic, objective ratings in a short time.

The Serviceability Tools and Methods (ST&M) Approach

The ST&M approach (Davis et al., 1993) includes the use of the ASTM standards, and its results, but also

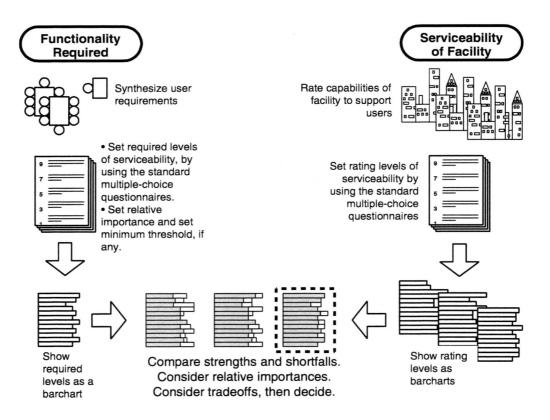

FIGURE A-4 Matching demand and supply—gap analysis. Source: Francoise Szigeti and Gerald Davis, © 1999, 2000 International Centre for Facilities.

provides formats for describing the organization, function-based tools for estimating how much floor area an organization needs, and other tools necessary to provide needed information for the statement of requirements (SOR).

At the heart of this approach is the process of *working with* the occupant groups during the programming phase of the project cycle, as well as during any evaluation phase. This process of communication between the providers of services and products (in-house and external) and the other stakeholders (in particular the occupants), of valuing their input, and of being seen to be responsive can be as important as the outcome itself and will often determine the acceptability of the results.

This is where satisfaction and quality overlap. ST&M includes several kinds of methods and tools, along with documents and computer templates for using them:

1. Functional requirement bar-chart profile and functional elements

2. Facility serviceability bar-chart profile and indicators of capability
3. A match between two profiles and comparisons with up to three profiles
4. A gap analysis and selection of "strength and concerns" for presentation to senior management
5. Text profiles for use in a statement of requirements and equivalent indicators of capability
6. Descriptive text about the organization, its mission, relevant strategic information, and other information about the project in a standard format
7. Quantity spreadsheet profiles
8. Building loss features (BLF) rating table
9. Footprint and layout guide

The ASTM standards and the ST&M approach are project independent. Requirements profiles can be prepared at any time, and serviceability ratings can be done and updated at a number of points over the life cycle of a facility.

How Do These Tools Fit in the Overall Corporate Real Estate Framework?

Setting requirements and evaluating results are two parts of what should be an ongoing dialogue between users and providers. Evaluations are becoming an indispensable tool for decision-making by senior management and for appropriate responsiveness by all those involved, whether they are working in-house or are external providers.

Facilities are an important resource of the enterprise. There are three main processes to take into account: (1) demand; (2) management, planning, procurement, production, and delivery; and (3) operations, maintenance, and use.

Figure A-5 shows how the different processes involved relate to each other and to the enterprise.

In Figure A-6, the functionality and serviceability scales are shown as an overlay on this framework. In this manner, it is possible to see how they relate to the underlying corporate real estate processes. The processes diagrammed here are each complex sets of activities and secondary processes. A detailed map of such activities is included in the next volume of scales to be published by the ICF (Davis et al., in press).

Figure A-6 also shows the relationship to a new set of scales prepared to rate the condition and estimated residual service life of a facility, to compare them to the needs of the enterprise. These scales are also used

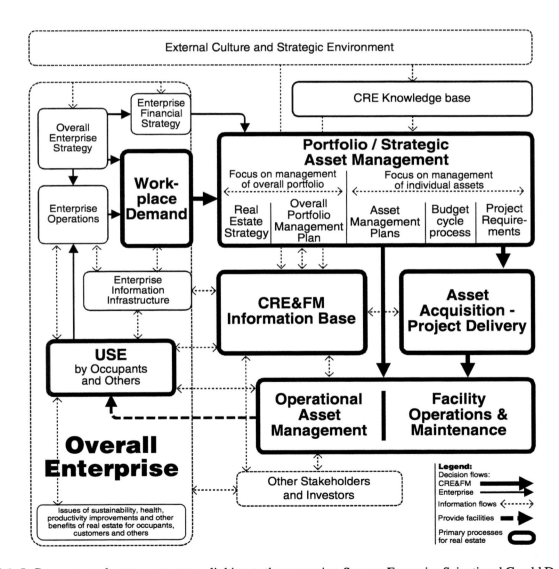

FIGURE A-5 Corporate real estate processes—linking to the enterprise. Source: Francoise Szigeti and Gerald Davis, © 1999, 2000 International Centre for Facilities.

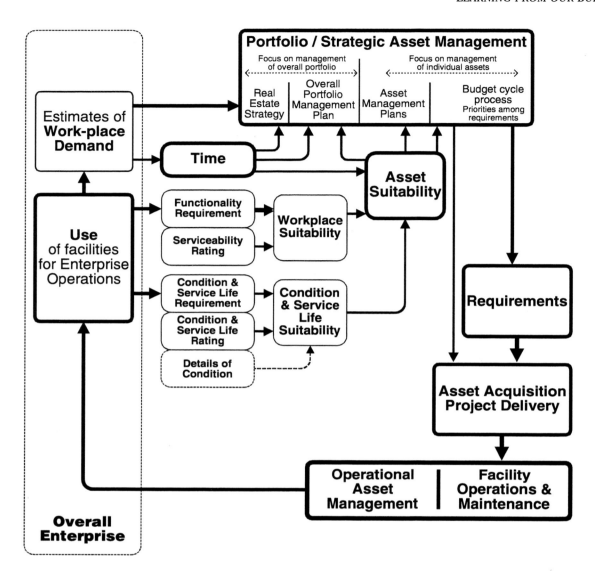

FIGURE A-6 Corporate real estate processes and use of the serviceability tools. Source: Francoise Szigeti and Gerald Davis, © 1999, 2000 International Centre for Facilities.

for setting budget priorities for repair and alteration projects.

EXAMPLES: USES OF THE ASTM STANDARDS AND LINKS TO OTHER TOOLS

Link to GSA's Customer Satisfaction Survey: GSA, John C. Kluczynski Building, Chicago

GSA regularly assesses the satisfaction of occupants of its major buildings, using a version of the customer satisfaction survey developed by the International Facility Management Association. The satisfaction levels of occupants of its landmark John C. Kluczynski Building in downtown Chicago were compared to levels in a serviceability rating of the building and to functionality requirement profiles for the main categories of occupant groups.

The results correlated closely. The serviceability levels both predicted and explained the satisfaction levels. However, the customer satisfaction survey had more detail about how occupants felt about the speed and thoroughness with which building operations staff responded to problems and complaints. The serviceability scales gave more information about actual strengths and concerns of the facility to meet occupant functional needs.

Together, these two complementary studies provided needed supporting information to submissions for the funding of several renovation projects. An example of a bonus from the functionality and serviceability project was the identification of ongoing security concerns of the staff. Occupants did not realize that the situation had been remedied and that thefts had been reduced by three-quarters since the assignment of a policeman in uniform on-site, patrolling the corridors. On the basis of this particular finding, a communications and public relations effort was launched to inform the staff of the beneficial impact of the presence of the uniformed policeman.

Link to Prior POEs: U.S. State Department

During the 1990s, the U.S. State Department conducted POEs after many major projects. The findings from these projects were analyzed by in-house staff and others of the Office of Foreign Buildings Operations (now Office of Overseas Buildings Operations). Then, in the late 1990s, a functionality requirement profile was developed for chanceries, using the ASTM standard scales. Data, as well as insights, from the POEs were taken into account in setting requirement levels and in preparing the main requirement profile for a base building as well as for the variant profiles for the different zones in a chancery. The ASTM standard scales provided a structure for applying the information from the POEs, which could then be directly related to equivalent levels of serviceability.

Part of Asset Management Plans: Public Works and Government Services Canada

When the serviceability tools and methods were first developed, the government of Canada rated the serviceability of all its major office buildings across the country. Recently, it has issued contracts to update the asset management plans of all its major office buildings. Each plan is required to contain a serviceability rating using the standard serviceability scales.

Part of Tool Kit for Portfolio Management and Setting Budget Priorities for Repair and Alteration Projects: PG&E

Pacific Gas & Electric Corporation (PG&E) has realigned the way it plans and manages its portfolio of real estate and sets priorities in its annual budget for repairs and alterations. Integral to its new management approach are the ASTM standard scales, the serviceability tools and methods including the new scales for building condition, and estimated remaining service life. These ratings are compared to the required level set for that facility.

New Design: National Oceanic and Atmospheric Administration (NOAA)

The groups that operate the weather satellites of NOAA needed a new headquarters building. Their functional requirements were specified using the ASTM standard scales. Their requirement profile was then compared to that of private sector organizations doing similar work, such as the headquarters of a gas pipeline company or the headquarters of a mobile phone company. NOAA's requirement profile was very similar to what was needed by other organizations doing similar kinds of work; there were very few differences. This provided a kind of benchmarking for NOAA's senior management and showed that NOAA's requirements were appropriate and consistent with private sector practice, even though they were much more demanding than would typically be provided for a general administrative office in government.

Choosing a Lease Property: U.S. State Department, Passport Office

The U.S. State Department and GSA reached agreement on the functionality requirement profile for its passport offices where citizens can come to have their applications adjudicated and a passport issued quickly. When new leased office space was needed for its passport office in New Orleans, the requirement profile was verified for its applicability to this particular office. Then about a quarter of the requirement scales were used to scan the properties on offer that GSA had identified as relevant. In two days, six properties were scanned for serviceability levels. Only one out of six was found to meet the essential functionality requirements. However, once the real estate manager in GSA understood the requirement profile of the occupant group, she was quickly able to identify two other valid options. Within weeks, the "best fit" was identified, and a lease negotiated.

Requests for Proposals (RFPs) and Design Reviews: State Department Chanceries and High-Tech Organizations

During 1999 and 2000, the U.S. State Department used its functionality requirement profile and serviceability indicators to assess the functionality of design proposals for new embassies and consulates to be developed in the sequential project development process, design-bid-build. The department's requirement profile was also used to assess and compare the functionality of proposals using the integrated, design-build procurement process. In some cases, proposals had very similar levels of functionality, while some other proposals showed significant differences between the functionality of the proposals and what had been required.

Thereafter, the same requirement profile was used during design reviews as a benchmark to ensure that the designs were continuing to respond to the requirements stated in the RFP. Contractors were trained to do these assessments, to create comparisons bar-charts, and to analyze the gaps between the designs and the requirement profile.

When a slow-growth, high-tech corporation needed a new corporate headquarters, it developed a functional requirement profile in the language and format of the ASTM standards and included it, verbatim, in its RFP. Responses to the RFP were rated, using the serviceability scales. Although a number of the proposals were fairly tightly clustered on price, there was a significant difference in functionality among the proposals.

Levels of Service for Outsourcing: A Major Oil Company

When a major oil company was considering outsourcing its facility management operations, it asked all companies who proposed to base their cost proposals on using the same levels of serviceability, as specified using the ASTM standard serviceability scales. Senior management also asked the "in-house provider" to self-rate using the same standards. Both the in-house facilities group and the bidders were able to do this on short notice, without having received any special training or guidance in using the standards. Comparable proposals, based on these consistent requirements, were received on schedule without difficulty. This company also asked TEAG to rate its main campus and to prepare a profile of requirement for the largest occupant group housed at that campus. This third-party assessment served as a benchmark to compare the results from the other assessments and proposals.

FINAL COMMENTS

Some Relevant Anecdotes

The *value* of a property and its long-term benefits are not just a matter of real estate dollars and cents and technical performance; corporations also look at the effectiveness of the workplace for core business operations and at the strategic advantage that facilities can provide. Successful facilities and real estate groups understand this (FFC, 1998). It can cost or earn the company far more than the rise or fall of property prices. For example, one vice president for facilities at an aircraft company explained to us that facilities costs represent about 5 percent of the total cost of each airplane sold, but that 5 percent is critical to the ability of his company to deliver new planes on time and on budget. If a new hangar is not ready on time or the facilities get in the way, the whole production line can be delayed or grind to a halt. The same holds true for smaller companies who rent office space. For them the cost of rent, utilities, and other charges runs at about the same percentage.

For organizations, big or small, a 1 percent increase (or decrease) in the productivity of core business operations, brought about by an inadequate workplace, is probably at least 10 times greater than a 1 percent increase (or decrease) in the value of the real property considered as a real estate asset. Put another way, here is the example of a laboratory where facilities were underused and inefficient: At zero facilities cost and with minimum rearrangements, an extra 15 scientists could be added. This would give the lab extra gross revenue while lowering the cost of square feet per employee. On the other hand, more substantial changes in the facility layout would allow the lab to nearly double its population. This retrofit would cost far less than the cost of a new facility. Other functional improvements would increase the effectiveness of the staff by about 15 percent and speed time to market by about six months. All of these proposed changes were based on an assessment of functional capability. Overall, the asset value in use was expected to increase by more than $3,000,000, after taking into account retrofit costs. The earning power of the lab would be multiplied by almost 2.5. These numbers were calculated by one of the major accounting firms. Thus, typically, the greatest leverage for a facility comes from enhancing the performance of the core business. A factor of 10 or more is not unusual.

The effect of a facility on the health of its occupants can have a severe impact on productivity. Medical records are seldom used to prove this point but probably could be used more often. On one occasion, ICF was allowed to use records of sick days as part of a comprehensive facility evaluation after a major consolidation of staff into a single facility. After plotting the sick leave information for each of 18 groups for the year prior to the move, the year of the move, and the first year after the move, it became apparent that for all but two groups, the curve shifted up. For two groups, the curve shifted down. For those two groups, the building they came from was worse than the new facility. It was estimated that the number of days lost to the increased sick leave and a few other facility-related factors amounted to more than the annualized first cost of the building.

Sometimes, the effect of the facility can be drastic and immediate. In one case, due to some work being done in one part of the facility, traffic was redirected along an internal corridor cutting through the "territory" of a work group. What had been a "private path" was transformed overnight into a "major highway" (Davis and Altman, 1976). The partitions around that group were glass above 1 meter, which allowed passers-by to see into the space. ICF had warned that such a situation should not be allowed to happen because of the "fish-bowl" effect. The group in question was working on a critical path product that was at the heart of the future of the company and still highly secret. The group simply stopped work and did not put work on their desks. When the ICF team arrived on-site that day, it was asked to come directly to the office of the senior manager responsible for facilities. A work crew was commandeered to work overnight. Butcher paper was pasted over the glass to create visual privacy. By the next morning the problem had been solved, and work resumed.

In another case, staff retention was the victim. A major industrial corporation was recruiting young engineers to replenish its aging population, but these were leaving the company in record numbers after three months or less on the job. The human resources department conducted exit interviews to find out what the problem was. The young recruits reported that the space they were asked to work in was so unpleasant and antiquated that they felt the company had no regard for them. The job market at that time was good, and they could get jobs at other companies offering comparable salaries and much more attractive and modern facilities. This caused the company to start a $300 million rehabilitation program of its offices and to generally pay more attention to the physical setting of work.

Facilities have an impact when attracting staff, which can be the reverse of the last anecdote. Another major industrial company reported that it was located in an industrial area with other competitors. Its policy was to make its grounds attractively landscaped and to provide each of its software engineers with a private, well-furnished office with a window overlooking trees and flowers. The human resources department at that company could confirm that this "perk" was worth about 10 percent of payroll, that is, more than the annualized cost of the buildings and grounds.

Current Developments and Trends

Scales for Rating Condition and Estimated Remaining Service Life

As stated earlier, new scales have been developed by the ICF to enable a manager to set priorities for repair and alteration projects in the annual budget cycle. These scales are used for *building condition, estimated service life, and asset management.* They have been designed to assist managers to take into account the actual and required physical condition and estimated remaining service life of a facility or of its main systems and components. Typically, building condition reports give cost estimates to return a building to its original design but do not link directly to the level of functional support now required for occupant operations. These new scales can be matched directly, for gap analysis against the condition requirement profile in the asset management plan and overall portfolio management plan. They complement the information about the functional suitability of the facility to support the mission of the occupant group.

Scales for Other Building Types

A set of scales, similar to the ASTM standards, has been drafted for low-income housing and is being tested in New Zealand (Gray, 2001).

ICF has just completed additional scales for *service yards and maintenance shops*. Scales are also being developed to better cover sustainable building, manufacturing, retail, laboratories, education, health care, courts, and so forth.

Integrated Tools for Performance-Based Procurement

The Dutch government has mandated that all public procurement be performance based. With the ASTM standards as a starting point, the Dutch Building Agency has developed a systematic approach to define client expectations for total building performance (Ang et al., 2001). This approach also relates the translation of "inputs" and "outputs" at different phase of the project delivery process.

Strategic Asset Management

In some countries, portfolio management and strategic planning come under the term "strategic asset management" (SAM). In these places, asset is the preferred term, rather than facilities or buildings, of public sector managers who deal with all kinds of constructed capital assets, many of which may not be buildings. One of the most pertinent publications on the subject is a newsletter dedicated to the international review of all areas of performance and strategic management of assets, including economic considerations. Linking performance evaluations and costs is still a tricky business. Some of the concepts, such as profiles, levels of service, benchmarking targets, etc., which are becoming part of the performance evaluation, are explained in the newsletter in a practical and approachable way, with an emphasis on sharing of experience (Burns).

Levels of Service, Performance Profiles, Performance Benchmarks

The use of levels, or targets, is becoming prevalent for outsourced contracts, for performance-based procurements, and for strategic planning. The Department for Administrative and Information Services of the state of South Australia, is currently developing a building performance assessment and asset development

strategy, which adapts some of the ASTM standard scales and methodology to its own circumstances.

Warranty of Fitness for Purpose—Duty of Care Versus Duty of Results

In the wake of ISO 9000, with the advent of new integrated team approaches to projects, and performance-based procurement approaches, such as design-build, and construction manager at risk, there is an increased awareness of the need to state requirements more precisely and comprehensively and to be able to confirm that the resultant asset meets those requirements. Further, when delivering a full package, the contractor has the legal responsibility to deliver a product that fits the intended purpose. This is a major change for the traditional legal concepts of professional liability and duty of care, based on professional competence and accepted practice. This legal territory is being explored by groups such as the Design-Build Institute of America and the CIB (CIB, 1996).

In Conclusion

Evaluations are here to stay and will likely be taken for granted in the not too distant future. At a prior Federal Facilities Council (FFC) Forum, the presentation of the Amoco common process, developed by its Worldwide Engineering and Construction Division, included the following in Figure 2: "Operate—Evaluate asset to ensure performance . . . " (FFC, 1998).

To be more effective and useful, evaluations will have to be better coordinated with the information contained in statements of requirements. At the same FFC forum, several presenters included a project system in their presentation. The presentation by the director-construction of The Business Roundtable included a description of "Effective Project Systems" developed by the Independent Project Analysis Corporation of Reston, Virginia. He made the point that "the supply chain begins when the customer need is identified and translated into a business opportunity" (FFC, 1996). In such project systems, which usually are conducted by integrated project teams, the evaluation of alternative solutions is taken for granted. Thus, evaluations are part of the planning of the projects, not an afterthought.

The worldwide trend to deal with performance definitions rather than prescriptive or deemed-to-satisfy specifications will likely continue to spread. In the next few years, a further increase in the use of evaluations will be driven by the acceptance of performance based procurement by the World Trade Organization (WTO), the European Union, and the European member countries. Performance-based codes are also being adopted in a number of countries, including the United States.

The European Union (EU) is following suit. Since dealing with results rather than specifying solutions means that these results need to be shown to be performing as required, there is an assured future for evaluations and for evaluations as part of the process at many stages. Such developments, as they affect the building industry, will be the focus of a major thematic network being launched by the CIB, with EU funding (Bakens, 2001). This network will also include participants from the United States and other countries outside the EU. A key task will be how to prepare statements of requirements and their verification.

In the United States, a performance-based code has been adopted as a component of the new Unified International Building Code that has brought the three major codes together. Work is continuing at ASTM, the American Institute of Architects (AIA), NIST, and the GSA, to cite only a few of the key leaders. [4]

Benchmarking, lessons-learned programs, continuous improvements, and performance metrics are becoming part of business as usual. Indeed, to quote again from the 1998 FFC report: "What are the characteristics of the best capital project systems? In addition to using fully integrated cross-functional teams, they actively foster a business understanding of the capital project process. . . . The engineering and project managers are accountable to the business, not the plant management. There are continuous improvement efforts that are subject to real and effective metrics" (FFC, 1998). The evolution of POEs into building performance evaluations, and now into an ongoing evaluative stance, is likely to become the accepted norm because it is part of the best practices of companies that have succeeded in using capital projects in support of their primary business.

---

[4]ASTM Committee E06 on Performance of Buildings has oversight over this whole subject matter. The AIA has a Center for Building Performance. NIST continues to provide leadership with respect to housing and other more technical applications of the concept. The GSA, under the leadership of its Office of Governmentwide Policy, Real Estate, has embarked on a major program of research and publications on the subject.

ABOUT THE AUTHORS

Gerald Davis helps decision makers and facility managers implement solutions that enhance worker effectiveness, improve the management of portfolios of corporate real estate, solve facility-related problems, and ensure the optimum use of buildings and equipment. Since 1965, Mr. Davis has been considered a pioneer and internationally recognized expert in strategic facility planning, facility pre-design, programming, performance-based evaluations, and ergonomics. As senior author, he led the team that developed the *Serviceability Tools & Methods*® used to define workplace and facility requirements and to rate existing and proposed facilities. Previously, he led the "ORBIT-2" project, a major North American multisponsor study about offices, information technology and user organizations and about the impact of each on the other. He coauthored the "1987 IFMA Benchmark Report" which was the first of its kind. His work has been published in numerous trade and professional journals and books. He is the recipient of the Environmental Design Research Association Lifetime Career Award (1996), the IFMA Chairman's Citation, (1998), was named an IFMA fellow in 1999, and one of 50 most influential people in the construction industry by the *Ottawa Business Journal*. He is an ASTM fellow, and Certified Facility Manager, president and chief executive officer, International Centre for Facilities (ICF), president, TEAG (The Environmental Analysis Group), chair, ASTM Subcommittee E06.25 on Whole Buildings and Facilities; past chair, ASTM Committee E06 on Performance of Buildings; and past chair, IFMA Standards Committee (1993-99). He is also the U.S. (ANSI) voting delegate to the ISO Technical Committee 59 on Building Construction, to its Subcommittee 3 on Functional/User Requirements and Performance in Building Construction, and the former delegate to its Subcommittee 2 on Terminology and Harmonization of Language. Mr. Davis was recently appointed the Convenor of Work Group 14 on Functional Requirements/Serviceability.

Francoise Szigeti is the vice-president of the International Centre for Facilities, Inc., a scientific and educational public-service organization established to inform and help individuals and organizations improve the functionality, performance, and serviceability of facilities. She is also vice president and secretary-treasurer of The Environmental Analysis Group

(TEAG) - Groupe pour l'Etude du Milieu Humain (GEMH). She is president of Serviceability Tools & Methods, Inc. Ms. Szigeti is one of the vice-chairs of ASTM Subcommittee E06.25. She has served on the board of the Community Planning Association of Canada, the Environmental Design Research Association and the International Association for the Study of People and Their Physical Surroundings. She is a co-author and member of the team that developed the Serviceability Tools & Methods® used for defining workplace and facility requirements and for rating existing and proposed facilities. Previously, she launched and participated in the "ORBIT-2" project, a major North American multisponsor study about offices, information technology, and the user organizations, and about the impact of each on the other. She is a coauthor of the "1987 IFMA Benchmark Report", which was the first of its kind. Ms. Szigeti attended the Ecole Superieure d'Interpretes et de Traducteurs, Universite de Paris. She is the recipient of the EDRA Lifetime Career Award (1996).

REFERENCES

Ang, G., et al. (2001). *A Systematic Approach to Define Client Expectation to Total Building Performance During the Pre-Design Stage.* Proceedings of the CIB 2001 Triennial Congress.

ASTM (American Society for Testing and Materials). (2000). *ASTM Standards on Whole Building Functionality and Serviceability,* ASTM, West Conshohocken, Pa.

Baird, G., Gray, J., Isaacs, N., Kernohan, D., and McIndoe, G. (1996). *Building Evaluation Technique.* Wellington, New Zealand: McGraw Hill.

Bakens, W. (2001). Thematic Network PeBBu—Performance Based Building—Revised Workplan. Rotterdam:CIB.

Burns, P. (Ed.) *SAM—Strategic Asset Management Newsletter* AMQ International, Salisbury, South Australia.

CIB Publication 192. (1996). *A Model Post-Construction Liability and Insurance System* prepared under the supervision of CIB W087. Rotterdam, Holland.

Consumer Reports (2000). *Rating the Raters,* August 31.

Davis, G., and Altman, I. (1976). *Territories at the work-place: Theory into design guidelines.* In: Man-Environment Systems, Vol. 6-1, pp. 46-53. Also published, with minor changes, in: Korosec-Serfati, P. (Ed.) (1977). *Appropriation of Space, Proceedings of the Third International Architectural Psychology Conference* Strasbourg, France: Louis Pasteur University.

Davis, G., et al. (1993). *Serviceability Tools Manuals, Volume 1 & 2* International Centre for Facilities: Ottawa, Canada.

Davis, G. et al. (2001). *Serviceability Tools, Volume 3—Portfolio and Asset Management: Scales for Setting Requirements and for Rating the Condition and Forecast of Service Life of a Facility—Repair and Alteration (R&A) Projects.* International Centre for Facilities: Ottawa, Canada.

Federal Facilities Council. (1998). *Government/Industry Forum on Capital Facilities and Core Competencies.* Washington, D.C.: National Academy Press, p. 19.

Gibson, E.J. (1982). *Working with the Performance Approach in Building.* CIB Report, Publication 64. Rotterdam, Holland.

Gray, J. (in press). *Innovative, Affordable, and Sustainable Housing.* Proceedings of the CIB 2001 Triennial Congress. Rotterdam, Holland.

ISO 9000, Guidelines 9001 and 9004. (in process of reedition).

McGregor, W., and Then, D.S. (1999). *Facilities Management and the Business of Space.* Arnold, a member of the Hodder Headline Group.

National Bureau of Standards. (1971). *The PBS Performance Specification for Office Buildings,* prepared for the Office of Construction Management, Public Buildings Service, General Services Administration, by David B. Hattis and Thomas E. Ware of the Building Research Division, Institute for Applied Technology, National Bureau of Standards. Washington, D.C.: U.S. Department of Commerce NBS Report 10 527.

Szigeti, F., and Davis, G. (1997). Invited paper. In: Amiel, M.S., and Vischer, J.C., *Space Design and Management for Place Making.* Proceedings of the 28th Annual Conference of the Environmental Design Research Association (EDRA). Edmond, Okla.: EDRA.

Teicholz, E. (Ed.) (2001). *Facilities Management Handbook.* MacGraw-Hill.

Appendix B

A Balanced Scorecard Approach to Post-Occupancy Evaluation: Using the Tools of Business to Evaluate Facilities

Judith H. Heerwagen, Ph.D., J.H. Heerwagen and Associates

In the past decade, organizations have begun to look at their buildings not just as a way to house people but also as a way to fulfill strategic objectives (Becker and Steele, 1995; Horgen et al, 1999; Ouye and Bellas, 1999; Grantham, 2000). In part this is due to re-engineering and downsizing of the past two decades. Also, however, chief executive officers (CEOs) are beginning to think of their buildings as ways to achieve strategic goals such as customer integration, decreased time to market, increased innovation, attraction and retention of high-quality workers, and enhanced productivity of work groups.

Traditional post-occupancy evaluation (POE) methods do not provide the type of feedback needed to assess these organizational outcomes. POEs focus on individual-level assessment, most typically on satisfaction, use patterns, and comfort, rather than on organizational- or group-level outcomes associated with core business goals and objectives. Because organizations are increasingly asked to justify all of their major expenses, including facilities, evaluation methods that begin to address these higher-level issues would be of great value. At the present time, there are very few data to show linkages between facility design and business goals.

As a result, decisions are often made on the basis of reducing costs. Current cost-focused strategies include reducing the size of work stations, moving to a universal plan with only a few work station options, eliminating private offices or personally assigned spaces, and telecommuting. Evaluation methods that identify and measure the business value of facilities would be a highly valuable way to expand the current knowledge base and to provide a wider array of outcomes against which to measure facility effectiveness.

Ouye (1998; Ouye and Bellas, 1999) suggests that workspace design and evaluation can become more strategic by adopting the tools of business, specifically the Balanced Scorecard (BSC) approach proposed by Kaplan and Norton (1996). As applied to facilities, the BSC approach pioneered by Ouye and the Workplace Productivity Consortium is valuable not only for evaluation purposes, but also for design because it forces designers to think systematically about the relationship between the workplace and organizational effectiveness. Although the Balanced Scorecard was developed primarily with the private sector in mind, the approach is also applicable to the evaluation of government facilities. A core theme for both the private sector and the government is to provide facilities that are both efficient and effective. As noted in the General Services Administration's (GSA's) *The Integrated Workplace* (GSA, 1999, p.5):

> By using the Integrated Workplace as part of your strategic development plan, matching business goals to workplace designs, you can consolidate and reconfigure the spaces where you work while providing people with the tools they need to support the organization's mission.

Even more important for federal facilities is the strong link between the BSC approach and the requirements for strategic planning and performance assessment initiated by the Government Performance and Results Act of 1993 (GPRA). GPRA was enacted as part of the Clinton administration's "Reinventing Government" initiative to increase the efficiency of federal agencies and to make them more accountable for achieving program results. GPRA requires federal departments and agencies to develop methods for measuring their performance against strategic goals and program objectives, an approach that is very consistent with the Balanced Scorecard. Because the BSC focuses on evaluation as a means to enhance overall strategic performance, the results from the BSC approach would provide valuable input to the GPRA performance review for federal facilities. At the present time, measures of facility "success" include costs per square foot of space and square foot per occupant. Such measures do not address the strategic issues of concern to GPRA and the Balanced Scorecard.

The Balanced Scorecard assesses four categories of performance: financial, business process, customer relations, and human resource development (Kaplan and Norton call this dimension "learning and growth").

The term "balanced" refers to several factors. First, there is a balance across the four categories to avoid overemphasis on financial outcomes. Second, the evaluation includes both quantitative and qualitative measures to capture the full value of the design project. And third, there is a balance in the levels of analysis–from individual and group outcomes to higher-level organizational outcomes. Figure B-1 shows some possible measures to use in each category. All of these measures have logical links to the workplace environment, and in some cases there is empirical support also.

This appendix draws on the framework developed by Ouye (1998; Ouye and Bellas, 1999) but expands it to include greater discussion of the links between physical space and business outcomes. The approach and process described in this chapter also focus on dif-

FINANCIAL OUTCOMES	BUSINESS PROCESS OUTCOMES
• Operating/maintenance costs • Costs of building related litigation • Resale value of property • Rentability of space	• Process innovation • Work process efficiency • Product quality • Time to market
STAKEHOLDER RELATIONS	HUMAN RESOURCE DEVELOPMENT
• Public image and reputation • Customer satisfaction • Community relationships	• Quality of work life • Personal productivity • Psychological and social well being • Turnover • Cultural change

FIGURE B-1 Building evaluation measures using the BSC approach.

ferent categories of outcomes. Whereas Ouye's "workplace performance" process focuses on strategic performance, group performance, and workplace effectiveness, this appendix links evaluation more directly to the four dimensions of the Balanced Scorecard.

ADVANTAGES OF USING THE BALANCED SCORECARD APPROACH

At the present time, POEs focus on the human resource dimension of the BSC and primarily on quality of work life (which includes environmental satisfaction, comfort, functional effectiveness of space, access to resources, etc.) This is an important component, but only one indicator of the success of the facility. In fact, it is possible for a facility to rate very high on these characteristics, but to have a negative impact on the other areas if the building costs considerably more to operate and maintain or if the design interferes with key work processes in some way. This may happen if a design emphasizes visual openness to enhance communication at the expense of ability to concentrate (Brill and Weidemann, 1999).

The BSC approach, in contrast, begins by asking these kinds of questions:

- How can workplace design positively influence outcomes that organizations value?
- How can it reduce costs or increase income?
- How can it enhance human resource development?
- How can the workplace enhance work processes and reduce time to market?
- How can the work environment enhance customer relationships and present a more positive face to the public?

By asking these questions at the beginning of the design project, the BSC approach provides an analytical structure to the entire process, from conceptualization through evaluation and finally to "lessons learned." For an organization or design firm, these lessons learned become the knowledge base for future design efforts.

One of the trade-offs inherent in using the BSC approach, however, is the difficulty of generalizing to different contexts. Because the evaluation methodology is so closely linked to a unit's own mission and objectives, it is difficult to generalize findings to other spaces and units. To deal with this difficulty, a core set of measures could be used across facilities to gain a better

understanding of the entire building stock, while other measures would be unique to the goals and objectives of the particular organization, department, or division (Ouye, 1998).

APPLYING THE BSC APPROACH

In applying the BSC approach, the following general steps need to be taken:

1. Identify overall organization mission and specific objectives for each of the four BSC dimensions.
2. Identify how the facility design is expected to help meet each objective.
3. Select specific measures for each of the organizational objectives based on links to the workplace design. Set performance goals for each measure.
4. Conduct evaluation "pre" and "post."
5. Interpret findings in light of the mission and objectives.
6. Identify key lessons learned.

An example will help illustrate this process.

1. Identify Mission and Objectives

Mission: Become a showcase government office of the future.

Specific objectives for each BSC dimension:

- Financial—reduce the costs of modifying facilities.
- Business process—reduce time for delivery of products; create more collaborative working relationships within and between units
- Stakeholder relationships—upgrade the image of government workspace, increase customer satisfaction.
- Human resources—improve overall quality of work life; reduce turnover and absenteeism.

2. Identify Potential Links to the Physical Environment

This step of the evaluation process is the most neglected in facility evaluation. It requires conscious articulation of design hypotheses and assumptions about expected links between the specific objectives and the features of the environment. By making these potential links more explicit, it will be easier to interpret results and to assess differences between spaces that vary on key physical features and attributes. Furthermore, by testing specific design hypotheses, the BSC approach can be used to test and develop theory. In contrast, most POE research is theoretically weak and does not contribute to either hypothesis testing or theory development.

This step requires research on what is known already as well as logical speculations. The hypothesized links form the basis for characterizing the baseline and new environments. It includes physical measures of the ambient environment (thermal, lighting, acoustics, and air quality) and characterization of other key features and attributes of the environment that are known or suspected to influence the outcomes of interest to the study. The specific features and attributes used in the characterization profiles are related to the objectives and measures. For instance, if one of the objectives is to reduce absenteeism, then features of the environment known to influence illness need to be catalogued pre and post. These include materials selection for carpeting and furnishings; ventilation rates; ventilation distribution; thermal conditions; cleaning procedures; and heating, ventilation, and air conditioning (HVAC) maintenance (Fisk and Rosenfeld, 1997; Heerwagen, 2000). The profiles will enable the evaluation team to understand better what works and what does not and why. They can also be incorporated into a database that can be integrated across facilities.

For example, one of the measures cited above for improved work process is increased collaboration both within and between units, especially support for spontaneous interactions and meetings. A quick review of research literature provides the basis for developing the potential links to the environment. Specific questions to address in the literature review include: How do people use the environment for social interactions? What aspects of the environment encourage different kinds of interactions? How do groups work, and what resources do they typically use? How important are spontaneous meetings compared to planned meetings, and how do they differ from one another?

Recent research on informal communications and interactions in work settings shows specific features and attributes of informal spaces that are likely to influence the extent to which the spaces are used and their degree of usefulness to work groups. These features include comfortable seating to encourage lingering, location in areas adjacent to private workspaces to encourage casual teaming, white boards for discus-

sions, presence of food nearby, some degree of separation from main traffic routes, acoustical separation for nearby workers who are not participating, ability of team members to personalize the space, and ability to maintain information displays, group artifacts, and work in progress until these items are no longer needed (Allen, 1977; Sims et al., 1998; Brager et al., 2000). Once these potential features are identified, the baseline and new environments can be assessed to identify the extent to which these features are present.

A similar process would be carried out for each of the objectives. For many of the objectives, there is likely to be little research available. Nonetheless, the assumptions, hypotheses, and predictions should still be articulated and linkages to the environment should be logically consistent. As another example, reduced time to market could be influenced by factors such as co-location of people working on the task, easy access to electronic groupware tools to coordinate work, sufficient group space for spontaneous meetings, vertical surfaces for continual visual display of work in progress and schedules, and central storage for materials associated with the task while it is ongoing.

3. Identify Specific Measures for Each Objective

Many different kinds of measures are likely to be used in the evaluation process. Nonetheless, each potential measure should be assessed against general criteria to help decide whether or not it should be included in the evaluation process. These criteria follow:

- *Usefulness*—the measure addresses the mission, goals, and objectives of the business unit and can be used in strategic planning.
- *Reliability*—the measure produces consistent results when applied again.
- *Validity*—the measure is a good indicator of the outcome of interest (it measures what it purports to measure).
- *Efficiency*—the overall measurement plan uses the minimal set of measures needed to do the job and enables conclusions to be drawn from the entire data set.
- *Discrimination*—the measures will allow small changes to be noticed.
- *Balance*—the entire set of measures will include both quantitative and qualitative measures and direct and indirect measures. *Quantitative* data can be translated into numbers and used for

statistical analyses. *Qualitative* data, on the other hand, often include interviews and results from focus groups that are more difficult to translate into numeric scales. Nonetheless, such data provide a rich understanding of the context and processes that make it easier to interpret quantitative results. Further, qualitative approaches are often used as a means to develop items for surveys and structured interviews or other data-gathering mechanisms. The second aspect of a balanced family of measures is direct versus indirect measures of performance. *Direct* measures include outputs such as sales volume. *Indirect* measures are often correlated with performance or are the building blocks of performance rather than actual performance output. Examples are frequency of use, occupant satisfaction, or absenteeism.

Setting Performance Goals. The organization needs to decide prior to the evaluation what degree of improvement it is working toward for each of the identified measures. Does even the slightest increase in the expected outcomes matter? Alternatively, should you aim for a 10 percent improvement, a 25 percent improvement? Setting performance guidelines will help in data interpretation and conclusions. Scientific research uses statistical significance as proof of success. However, statistical analyses may not be as useful in an applied context. The degree and direction of change over time may be more relevant to organizational performance. Very few performance evaluation processes, including the Balanced Scorecard, use statistical analyses to judge whether organizational changes are "working." Instead, managers look at the overall profile of outcomes and make a decision about new policies or procedures based on how well the data match improvement goals (Kaplan and Norton, 1996).

The following examples identify some potential measures for each of the stated objectives in our hypothetical example.

Costs of Modifying Facilities. This would involve identification and calculation of all costs involved in relocating workers or reconfiguring office space, including costs associated with packing or unpacking, time and costs of facilities staff needed to reconfigure work stations, time associated with planning the move and reporting, costs for any special equipment or services, and lost work time. The data would include total costs of staff time (number of hours × hourly pay rate),

total costs of special equipment or services, and the amount of time overall to carry out the change from the initial request to completion. In order to demonstrate reduced costs (the objective), pre and post comparisons would focus on the total number of people involved in a move and the costs of their time; the overall time needed to carry out the change; and the total dollar costs of the move.

Delivery Time for Products. This would require tracking the time to actually produce a product such as a report, starting with the initial assignment and ending with the date of final delivery. To identify where in the overall process the efficiencies occurred, subtasks would also be timed. Data would include a written commentary on work process, number of people involved in the task and their responsibilities, and reasons for any unusual delays or work stoppages. The best tasks for such an analysis are reports or other products that are done on a repetitive basis and therefore are likely to be very similar from year to year. If different types of products are selected pre and post, then any differences in delivery times could be due to factors such as task complexity rather than to increased efficiencies resulting from changes in physical space.

Facility Image. Data on image would include subjective assessments through brief surveys completed by visitors, customers, job applicants, and staff. Specific questions would depend upon the nature of the facility and the type of work. Pre-post analysis would look for changes in perceptions.

Customer Satisfaction. Data on customer satisfaction could include surveys, analysis of unsolicited customer messages (complaints, concerns, praise), customer retention, and number of new customers. Objective data could include the time needed to complete transactions with different customers or stakeholders or the number of requests for information processed per day.

Inter- and Intra-unit Collaboration. Communications and collaboration activities are notoriously difficult to document accurately unless logs are kept of all meetings, formal and informal. Furthermore, the value of collaboration is reflected not only in the frequency of the meetings, but also in the outcomes of the interactions (e.g., new ideas, problems solved more quickly). To get as accurate a picture as possible of changes in meeting characteristics, multiple methods should be used. First, motion sensors can be used to gather data on frequency of use for particular spaces. The sensors would have to contain counters or other data-processing technology (e.g., a sensor that would measure how long the lights were on) that would ignore short-duration changes (e.g., someone walking into the room briefly and then leaving). The occupancy data would have to be supplemented either with behavioral observations or surveys and interviews that gathered information on number and character of meetings attended within the past week (or some other limited time period to enhance the potential for accurate recall). The survey-interview process would also gather data on the attendees, the nature of the meeting (spontaneous versus planned; focused on a specific problem, brainstorming, task integration, information exchange, and so forth), and the perceived value of the meeting (specific outcomes, usefulness, etc.).

Data analysis would compare the number of meetings, the participants (number from within the unit, number from other units), the purpose, the outcome, and the perceived value. If the facility had an impact on collaboration, one would expect to find a wider range of participants, more meetings for problem-solving and brainstorming versus simple information exchange, more spontaneous meetings, and a higher perceived value.

Quality of Work Life. POEs traditionally focus on quality-of-work-life issues such as comfort, environmental satisfaction, work experiences and perceptions, sense of place, and sense of belonging. Many design firms and research organizations have examples of surveys that are used in a pre-post analysis.

Turnover. There is a great interest in retaining workers due to the high costs of turnover, in terms of both the financial costs associated with hiring someone new and the knowledge costs that result from losing valuable skills and knowledge when a worker leaves. Turnover is usually calculated as a rate of workers who voluntarily leave an organization divided by the total number of workers for the same time period. Turnover does not include retirements, dismissals, deaths, or loss of staff due to disabling illness. Some degree of staff turnover is important because it introduces new ideas and new skills into an organization. Thus, for evaluation purposes, the organization needs to decide what degree of turnover is desirable. Pre-post comparisons

of turnover rates should then be assessed against what is considered a desirable level.

4. Conduct Evaluation Pre and Post

Key issues in conducting the pre-post process include gaining cooperation from managers and staff who will be the study subjects, timing of the evaluation, and use of control or comparison groups.

Gaining Cooperation. The evaluation process will fail if the occupants are reluctant to participate or if there are insufficient staff to help with the organizational data gathering for some of the measures (such as turnover rates). Occupants are much more likely to continue to be engaged in the process if they are involved in helping design the measurement plan and if they see a benefit from participation. Having support from high-level organizational leaders is also critical because it signals the importance of the project. The facility occupants also need to be informed of how the data will be used and they need to be assured that their own input will be kept confidential.

Control Groups. Because so many other factors can influence the outcomes being studied, it is difficult to know whether performance changes are due to the workplace itself or to other factors that may change simultaneously. This is especially true when the design is part of an organizational change effort, which is often the case. Confounding factors may be internal to the organization (changes in policies or markets), or they may be external to the organization but nonetheless can affect business performance (such as economic conditions). The best way to avoid problems of interpreting the success of a design is to use control groups along with pre-post studies. An appropriate control group would be a business unit in the same building that does a similar kind of work but is not going through a workplace change. The control group should be as similar to the design change group as possible.

The control group is studied at the same time as the group experiencing the design changes, with both groups studied during the pre and post design phases. Although the control group would not experience the design change, it would get the same surveys or other measures at the same time. If the design has an impact independent of organizational issues, then the pre-post responses for those in the design change condition should show greater differences across time than for those in the control group.

Timing of Measurement. The pre-measurement should be done at least two to four months prior to the move into a new facility in order to avoid issues and problems associated with the move itself. Ideally, the existing facility should be evaluated before work begins on the new facility, although this is very seldom done due to the need to assemble a research team and develop a measurement plan. The post-measures should be done six to nine months after project completion to enable workers to adapt to the new setting. The delay will help to diminish the "settling-in" phase when problems may be most obvious and the workplace needs to be fine-tuned. It will also reduce the impact of a "halo" effect associated with being in a new or renovated space.

5. Interpreting Results

When data analysis is complete, the project reconsiders the design hypotheses and asks: Do the data support the hypotheses? Do the results meet the performance goals?

There are very few scientific research studies that show complete support for all hypotheses and predictions. Thus, we would not expect to find perfect alignment in design evaluation. Where misalignments occur, it is important to try to understand why this happened.

The design and evaluation teams will have a natural tendency to focus on the positive and ignore the results that do not turn out the way they expected. However, it is often more valuable to understand why things went wrong for several reasons. First, you do not want to repeat the mistakes. Second, negative results often force a rethinking of basic assumptions and a search for better links between the environment and the behavioral outcomes.

A problem with all facility evaluations, regardless of specific methodologies, is the issue of causation. If a new facility is found to have positive outcomes, can these be attributed unequivocally to the physical environment and not to other factors? The answer is clearly no. Causation can be determined only by carefully designed experiments that vary only one component at a time. Since this is unrealistic in field settings, the causation issue will always be present. The best that can be done is to minimize other explanations to the degree possible through the use of control and comparison groups. It will also be important to use a high degree of logic in interpreting results, to look for consistency across facilities that share similar features, and to look at relationships between measures.

For instance, if absenteeism is of interest, then absenteeism rates should be associated with other outcomes, such as symptom expression or low levels of motivation, both of which could lead to taking days off due to illness or lack of desire to come to work. Assessing patterns of absenteeism will further aid in interpretation of results. Absence associated with motivational issues is likely to have a different pattern of days off than absenteeism due to illness. Because illnesses happen randomly and often last for more than one day, absenteeism due to illness should be clustered and randomly distributed over the days of the week. Motivational absenteeism, on the other hand, is more likely to occur on particular days of the week (especially Monday or Friday) and would be more likely to occur just one day at a time, not for several days. Another way to assess absenteeism is to look at its opposite—attendance. Attendance can be assessed by determining the percentage of workers with perfect attendance or the percentage who used less than the allowed number of sick days in the year prior to and the year after the move to the new facility. In addition to looking at relationships between measured outcomes, there should also be a logical connection to the physical environment profiles. Absenteeism and illness symptoms should be associated with factors such as poor indoor air quality and low maintenance of HVAC systems.

Another problem for interpreting the results on facility evaluation is that redesign often goes hand in hand with organizational change. Thus, positive (or negative) results could be due to organizational issues and not to the physical environment. This is where control groups become very valuable. If the organizational change is widespread, then similar units should also experience the effects of these changes. Thus, differences between the control group and the group in the new space are more likely to be related to the environment. Again, the use of logical thinking is also important. When organizational change occurs, some aspects of behavior are more likely to be influenced than others. For instance, if staff perceive the change very negatively, then motivationally influenced absenteeism may go up in both the new and the control spaces. Other outcomes, such as the costs of "churn," are less likely to be affected by organizational change policies.

Greater assurance of a true connection between the

physical features of the space and the measured outcomes can be gained also by using a geographical information system (GIS) approach to data analysis. Outcomes on various subjective measures can be plotted on floor plans to gain a greater understanding of the spatial distribution of responses. For instance, a GIS format used by the author to assess environmental satisfaction and comfort in an office building in California clearly showed that problems associated with distractions occurred primarily in particular locations, regardless of whether people were in private offices or cubicles. Similarly, thermal and air quality discomfort tended to cluster more in some areas than others. At the present time, most post-occupancy data analysis uses human characteristics as the primary unit of differentiation (e.g., different job categories, gender, age), with comparisons in responses across job categories or age. By supplementing the demographic data with geographical analysis, the evaluation will provide a more complete picture of the facility. A similar process, called spatial modeling, has been suggested by Aronoff and Kaplan (1995). Both the GIS and the spatial modeling approaches allow for analysis of the variability and distribution of responses in a spatial format.

7. Identifying Lessons Learned

An issue with lessons learned is: Where should the knowledge reside—in people's heads or in the environment? Should the lessons be internalized and become part of one's tacit operating knowledge, or should the lessons be located for anyone to access—in reports, databases, and so forth? Both should happen. If people are going to work with the knowledge gained, they need to incorporate it into their everyday ways of thinking and working. Internalization takes time and continued work with the knowledge and issues (Norman, 1993; Stewart, 1999). Once internalized, knowledge is part of a person's intellectual capital and leaves the organization when the person does. This is why knowledge also needs to be made explicit so it can become an organizational asset, not just a personal asset (Stewart, 1999). Seminars and presentations on the results of facility evaluations—with both positive and negative results highlighted—should be an ongoing practice.

Since a major purpose of evaluation is to apply the knowledge gained to future projects, simple databases that could be accessed by key words would be especially valuable to future designs. The database should include the design hypotheses and assumptions for each project, the specific measures used to test the hypothesis, pre and post photos of the space, brief summaries of the data, key lessons learned, connections to other studies, connections to the full research findings pre and post, and recommendations for future designs. The presentation of lessons learned should be as visual as possible to allow for maximum understanding and retention (Norman, 1993). Graphs, photos, and key words and concepts are much more likely to be useful than long verbal explanations that can be accessed if desired through links to full reports.

In addition, simple methods to display overall results would aid interpretation and lessons learned. For instance, results could be visualized using color-coded icons to provide an easy visual interpretation: green could be used to show strong support for the hypothesis and meeting or exceeding performance goals; yellow could be used to show minimal support or no change; and red could be used for measures that did not support the hypotheses or showed negative change.

SUMMARY AND CONCLUSIONS

This appendix has described an approach to post-occupancy evaluation that is more closely linked to business and workplace strategies than existing methodologies. Although the Balanced Scorecard approach does not present any new measurement techniques or breakthrough methodologies, it does provide a process for more effectively linking facilities to an organization's overall mission and goals. An advantage to using the BSC approach for federal facilities is its close relationship to the comprehensive performance assessment required by GPRA. Traditional POEs provide an important source of input, but measures tend to be focused on a limited range of topics and on the occupants' perspective, rather than on the broader, strategic focus of the BSC.

For large real estate portfolios, such as those in federal departments, the determinant of facility success is not only how well the overall building stock performs with respect to core POE measures used across facilities, but also how well each design fits its particular context and how well it meets the business objectives of the unit. The Balanced Scorecard was developed specifically for the purpose of providing data to assess overall performance and to identify areas that need attention.

ABOUT THE AUTHOR

Judith H. Heerwagen is an environmental psychologist whose research and writing have focused on the human factors of sustainable design and workplace ecology. Dr. Heerwagen currently has her own consulting and research practice in Seattle. Prior to starting her own business, Dr. Heerwagen was a principal with Space, LLC, a strategic planning and design firm, and a senior scientist at the Pacific Northwest National Laboratory. At Space Dr. Heerwagen was codirector of research and helped develop metrics for the Workplace Performance Diagnostic Tool. At the Pacific Northwest National Laboratory she was responsible for developing research methodologies to assess the human and organizational impacts of building design. Dr. Heerwagen has been an invited participant at a number of national meetings focused on workplace productivity. She was on the research faculty at the University of Washington in the College of Architecture and Urban Planning and at the College of Nursing. Dr. Heerwagen is a member of the American Psychological Association. She holds a bachelor of science in communications from the University of Illinois, Champaign-Urbana, and a Ph.D. in psychology from the University of Washington.

REFERENCES

Allen, T. (1977). *Managing the Flow of Technology*. Cambridge, Mass.: MIT Press.

Aronoff, S., and Kaplan, A. (1995). *Total Workplace Performance: Rethinking the Office Environment*. Ottawa: WDL Publications.

Becker, F., and Steele, F. (1995). *Workplace by Design: Mapping the High Performance Workscape*. San Francisco, Calif.: Jossey-Bass.

Brager, G., Heerwagen, J., Bauman, F., Huizenga, C., Powell, K., Ruland, A. and Ring, E. (2000). *Team Spaces and Collaboration: Links to the Environment*. Berkeley: University of California, Center for the Built Environment.

Brill, M., and Weidemann, S. (1999). Workshop presented at the Alt.Office99 Conference, San Francisco, Calif. December.

Fisk, W., and Rosenfeld, A.H. (1997). Estimates of improved productivity and health from better indoor environments. *Indoor Air* 7:158-172.

General Services Administration. (1999). *The Integrated Workplace: A Comprehensive Approach to Developing Workspace*. Washington, D.C.: Office of Governmentwide Policy and Office of Real Property.

Grantham, C. (2000). *The Future of Work: The Promise of the New Digital Work Society*. New York: McGraw-Hill, Commerce Net Press.

Heerwagen, J. (2000). Green buildings, organizational success and occupant productivity. *Building Research and Information* 28(5/6):353-367.

Horgen, T.H., Joroff, M.L., Porter, W.L., and Schon, D.A. (1999). *Excellence by Design: Transforming Workplace and Work Practice*. New York: Wiley.

Kaplan, R.S., and Norton, D.P. (1996). *The Balanced Scorecard*. Boston: Harvard Business School Press.

Norman, D. (1993). *Things That Make Us Smart: Defending Human Attributes in the Age of the Machine*. Reading, Mass.: Addison-Wesley.

Ouye, J.O. (1998). Measuring workplace performance: Or, yes, Virginia, you *can* measure workplace performance. Paper presented at the AIA Conference on Highly Effective Facilities, Cincinnati, Ohio, March 12-14.

Ouye, J.O., and Bellas, J. (1999). *The Competitive Workplace*. Tokyo: Kokuyo (fully translated in English and Japanese).

Sims, W.R., Joroff, M., and Becker, F. (1998). *Teamspace Strategies: Creating and Managing Environments to Support High Performance Teamwork*. Atlanta: IDRC Foundation.

Stewart, T.A. (1999). *Intellectual Capital*. New York: Doubleday.

Appendix C

Supplemental Information to Chapter 3

Buildings-in-Use

271 Lincoln Street
Lexington
MA.02421
781-674-3186 phone
781-674-1489 fax

Buildings-In-Use

Boston   Montreal

December 20, 2001

I.D.Number ____/ ___/ ____

Please leave blank

Welcome to the Building-In-Use Assessment Survey!

This questionnaire is for all staff. We want to find out more about how you feel about the facility you work in, and how you feel this environment affects your work.

Below you will find a checklist of items about your workspace. Please answer these questions as soon as you receive the questionnaire. It will take you less than 10 minutes to complete. **When you have filled it out, please return it immediately**.

Please do not fill out the ID number on this survey form. It is used for analysis purposes. However, please provide your office location in the space provided, as this will help us understand the building conditions at your work location. Your name is not necessary on the questionnaire and your answers will remain confidential.

We really want to hear from you. Thank-you for participating!

PLEASE FILL OUT THE FOLLOWING: **Office or cube number** _____

Floor _____   **Workgroup or department name** _____

Please rate your comfort level in your primary workspace on the following scales, where **1** is poor or uncomfortable and **5** is good or comfortable, and **2** ☐**3 - 4** are in-between, with **3** being neutral. **Your task is to circle the number on each scale that best represents your experience of working in this building.**

1. Temperature comfort:

1	2	3	4	5
GENERALLY BAD				GENERALLY GOOD

over . . .

1

Buildings-in-Use

2. How cold it gets:

1	2	3	4	5
TOO COLD				COMFORTABLE

3. How warm it gets:

1	2	3	4	5
TOO WARM				COMFORTABLE

4. Temperature shifts:

1	2	3	4	5
TOO FREQUENT				GENERALLY CONSTANT

5. Ventilation comfort:

1	2	3	4	5
GENERALLY BAD				GENERALLY GOOD

6. Air freshness:

1	2	3	4	5
STALE AIR				FRESH AIR

7. Air Movement:

1	2	3	4	5
STUFFY				CIRCULATING

8. Noise distractions:

1	2	3	4	5
DISTURBING				NOT A PROBLEM

9. General office noise level (background noise from conversation and equipment):

1	2	3	4	5
TOO NOISY				COMFORTABLE

10. Specific office noises (individual voices and equipment):

1	2	3	4	5
DISTURBING				NOT A PROBLEM

over . . .

2

Buildings-in-Use

11. Noise from the air systems:

1	2	3	4	5
DISTURBING				NOT A PROBLEM

12. Noise from office lighting:

1	2	3	4	5
BUZZ/NOISY				NOT A PROBLEM

13. Noise from outside the building:

1	2	3	4	5
DISTURBING				NOT A PROBLEM

14. Furniture arrangement in your workspace:

1	2	3	4	5
UNCOMFORTABLE				COMFORTABLE

15. Amount of space in your workspace:

1	2	3	4	5
INSUFFICIENT				ADEQUATE

16. Work storage:

1	2	3	4	5
INSUFFICIENT				ADEQUATE

17. Shared (team) file storage:

1	2	3	4	5
INSUFFICIENT				ADEQUATE

18. Personal storage:

1	2	3	4	5
INSUFFICIENT				ADEQUATE

19. Visual privacy in your workspace:

1	2	3	4	5
UNCOMFORTABLE				COMFORTABLE

over . . .

3

Buildings-in-Use

20. Voice privacy in your workspace:

1	2	3	4	5
UNCOMFORTABLE				COMFORTABLE

21. Telephone privacy in your workspace:

1	2	3	4	5
UNCOMFORTABLE				COMFORTABLE

22. Electrical Lighting:

1	2	3	4	5
UNCOMFORTABLE				COMFORTABLE

23. How bright lights are:

1	2	3	4	5
TOO MUCH LIGHT				DOES NOT GET TOO BRIGHT

24. Glare from lights or windows:

1	2	3	4	5
UNCOMFORTABLE				COMFORTABLE

25. Natural lighting from windows:

1	2	3	4	5
INSUFFICIENT LIGHT				GOOD NATURAL LIGHT

26. Not enough light:

1	2	3	4	5
TOO DARK				COMFORTABLE

27. Please rate how this space affects your ability to do your work:

1	2	3	4	5
MAKES IT DIFFICULT				MAKES IT EASY

28. How would you rate your satisfaction with this building?

1	2	3	4	5
DISSATISFIED				VERY SATISFIED

over . . .

4

Buildings-in-Use

PLEASE COMMENT:

29. What I **like best/find most useful** about this building as a place to work :

30. What I **dislike most/have most trouble** with in this building as a place to work :

Appendix D

Supplemental Information to Chapter 4

NAVAL FACILITIES ENGINEERING COMMAND

Reference: CUSTOMER QUALITY SURVEY

Dear Navy/Marine Corps team member:

Thank you for taking the time to complete this survey. We have developed this survey to help understand what is **important to you about the new facility** with which you are associated. The results will allow Naval Facilities Engineering Command to recognize what makes a high quality facility in the eyes of its customers and to focus better on areas that might need improvement.

All the information provided by you in this survey will be treated as <u>private and confidential</u> and used only for NAVFAC to review their planning, design, construction and maintenance criteria and management turnover processes for continuous improvement. **We do not want your name on the survey.**

We value your input to the survey -- please answer as fully as you can. Thanks, again!

<u>Note 1:</u>   *Questions have both positive and negative wording. Be careful.*

<u>Note 2:</u>   *All areas may not directly apply to you. Do the best you can.*

<u>Section I: About Yourself</u>
Please provide some information about yourself and your connection to this facility. Information in this section will be used to assist NAVFAC in sorting out which "customer" groups have common concerns about the facility planning, design, construction and turnover process.

NAVAL FACILITIES ENGINEERING COMMAND
CUSTOMER QUALITY SURVEY

Your **connection** to this facility (<u>**check the best one**</u>):

_____ I use this facility as living quarters.

_____ I use this facility as a workplace.

_____ I supervise or manage users of this facility.

_____ My main job is to maintain this facility.

_____ I supervise or manage maintainers of this facility.

_____ I use this facility only as a guest or customer - **not** employed here or living here.

_____ Other **(please explain):**

Your **involvement in the construction** of this facility (check **<u>as many</u>** as applicable):

_____ <u>Had no part</u> in the planning, design, construction or maintenance turnover of this facility.

_____ Participated in <u>planning phase</u> (before facility was funded).

_____ Participated in <u>design phase</u> (after facility was funded).

_____ Participated in <u>construction phase</u> (interacted with ROICC on construction issues or changes).

_____ Participated in <u>maintenance turnover</u> phase (turnover from ROICC after construction).

_____ <u>Received training in maintenance</u> of facility from ROICC/contractor or NAVFAC.

NAVAL FACILITIES ENGINEERING COMMAND
CUSTOMER QUALITY SURVEY

Section II: Functionality - How well the facility supports your mission.

Please indicate how strongly you agree or disagree with these statements by checking the corresponding blank.	Strongly Agree				Strongly Disagree	No Opinion/ Don't Know/ Doesn't Apply
	5	4	3	2	1	
Facility seems well suited to our mission.	—	—	—	—	—	—
Visitors in this facility can find their way around easily.	—	—	—	—	—	—
Installed equipment is **not** appropriate for this facility.	—	—	—	—	—	—
Kitchen is well suited to our needs.	—	—	—	—	—	—
Facility floor plan is compatible with our organization.	—	—	—	—	—	—
Furnishings make the spaces more pleasing to work in.	—	—	—	—	—	—
Telephone receptacles conveniently placed.	—	—	—	—	—	—
Facility supports our computer usage.	—	—	—	—	—	—
There are **not** enough electrical outlets for all the equipment we use.	—	—	—	—	—	—
Workspace-to-workspace movement is quick and easy. (When I need to go see somebody else in the facility, I can get there conveniently.)	—	—	—	—	—	—
Electrical capability can be expanded without major modification of facility.	—	—	—	—	—	—
Facility is flexible enough to meet changing needs.	—	—	—	—	—	—

Comments: _____

_____

_____

NAVAL FACILITIES ENGINEERING COMMAND
CUSTOMER QUALITY SURVEY

Section III: Environmental Issues

Please indicate how strongly you agree or disagree with these statements by checking the corresponding blank.

	Strongly Agree			Strongly Disagree		No Opinion/ Don't Know/
	5	4	3	2	1	Doesn't Apply
There is a problem with indoor air quality.	—	—	—	—	—	—
Hazardous Materials can be managed safely in this facility.	—	—	—	—	—	—
Trash collection is a problem inside this facility.	—	—	—	—	—	—
Storage of cleaning equipment and materials is **not** a problem.	—	—	—	—	—	—
It's hard to keep this facility looking squared away inside.	—	—	—	—	—	—
Facility orientation (way it faces on site) uses sun, shade and prevailing wind to best advantage.	—	—	—	—	—	—

Comments:_____

_____

_____

NAVAL FACILITIES ENGINEERING COMMAND
CUSTOMER QUALITY SURVEY

Section IV: Quality of Life in your facility

Please indicate how strongly you agree or disagree with these statements by checking the corresponding blank.

	Strongly Agree			Strongly Disagree		No Opinion Don't Know/ Doesn't Apply
	5	4	3	2	1	
Heating and air conditioning make facility comfortable to work in.	—	—	—	—	—	—
Facility is conveniently accessible for visitors.	—	—	—	—	—	—
Facility is conveniently accessible for occupants.	—	—	—	—	—	—
It is easy for disabled persons to get around in this facility.	—	—	—	—	—	—
Disabled persons can operate all necessary functions of facility.	—	—	—	—	—	—
Lighting in facility is adequate.	—	—	—	—	—	—
Spaces provide the <u>work</u> privacy we need.	—	—	—	—	—	—
This facility is too noisy.	—	—	—	—	—	—
Attention to detail in construction is evident.	—	—	—	—	—	—
Material finishes are appropriate to overall purpose of facility.	—	—	—	—	—	—
Little things, like doorknobs, switches, faucets, etc., do **not** seem to work or fit.	—	—	—	—	—	—

Comments: _____

_____

_____

NAVAL FACILITIES ENGINEERING COMMAND
CUSTOMER QUALITY SURVEY

Section V: Safety in your facility
Please indicate how strongly you agree or disagree with these statements by checking the corresponding blank.

	Strongly Agree				Strongly Disagree	No Opinion/ Don't Know/
	5	4	3	2	1	Doesn't Apply
Exterior lighting provides adequate security for users of facility.	—	—	—	—	—	—
Facility design enhances physical security.	—	—	—	—	—	—
Emergency exits are clearly marked and easily accessible.	—	—	—	—	—	—
Fire alarms are accessible.	—	—	—	—	—	—
Disabled persons will have trouble getting out of facility.	—	—	—	—	—	—
Design of facility enhances safe operating conditions.	—	—	—	—	—	—
Safety systems for occupational hazards are readily available.	—	—	—	—	—	—

Comments: _____

_____

_____

NAVAL FACILITIES ENGINEERING COMMAND
CUSTOMER QUALITY SURVEY

<u>Section VI: Appearance of your facility</u>
Please indicate how strongly you agree or disagree with these statements by checking the corresponding blank.

	Strongly Agree				Strongly Disagree	No Opinion/ Don't Know/ Doesn't Apply
	5	4	3	2	1	
Facility looks good.	—	—	—	—	—	—
Facility fits well with overall appearance of base (size, design and color).	—	—	—	—	—	—
Interior design enhances work environment.	—	—	—	—	—	—
Landscaping looks good.	—	—	—	—	—	—
Main entry is a pleasing, inviting way into facility.	—	—	—	—	—	—
This facility is of award caliber.	—	—	—	—	—	—

Comments: _____

_____

_____

NAVAL FACILITIES ENGINEERING COMMAND
CUSTOMER QUALITY SURVEY

Section VII: The Planning Process for your facility

Please indicate how strongly you agree or disagree with these statements by checking the corresponding blank.

	Strongly Agree				Strongly Disagree	No Opinion/ Don't Know/
	5	4	3	2	1	Doesn't Apply
Planned size was adequate for actual requirements.	—	—	—	—	—	—
Facility site was well planned to accommodate services like delivery and trash removal.	—	—	—	—	—	—
Location of facility was well planned to fit organization's mission.	—	—	—	—	—	—
Moving to our new facility has had a negative effect on working with organizations outside the facility.	—	—	—	—	—	—
Space criteria for designing facility matched requirements of mission.	—	—	—	—	—	—
Parking is adequate, convenient and safe.	—	—	—	—	—	—
Mission changes since planning facility have made design inadequate.	—	—	—	—	—	—
Customer actively participated in planning process for new facility.	—	—	—	—	—	—

Comments: _____

_____

_____

NAVAL FACILITIES ENGINEERING COMMAND
CUSTOMER QUALITY SURVEY

Section VIII: Maintenance of your facility

Please indicate how strongly you agree or disagree with these statements by checking the corresponding blank.

	Strongly Agree				Strongly Disagree	No Opinion/ Don't Know/ Doesn't Apply
	5	4	3	2	1	
Roof has a problem with leaks.	—	—	—	—	—	—
Windows seal tightly against weather.	—	—	—	—	—	—
Ventilation system is quiet.	—	—	—	—	—	—
Air conditioning ducts drip.	—	—	—	—	—	—
Doors operate smoothly.	—	—	—	—	—	—
Windows operate smoothly.	—	—	—	—	—	—
Training received in maintaining this facility was about right.	—	—	—	—	—	—
Heating and air conditioning are too hard to operate.	—	—	—	—	—	—
We know what to do when something goes wrong with heating or air conditioning.	—	—	—	—	—	—
Plumbing works well.	—	—	—	—	—	—
Manuals received are clear and useful in maintaining facility systems.	—	—	—	—	—	—
Equipment is easy to access.	—	—	—	—	—	—
We put in a trouble call frequently on this facility.	—	—	—	—	—	—
Facility contractor did a good job of responding to problems.	—	—	—	—	—	—
Material finishes are easy to maintain.	—	—	—	—	—	—
Facility grounds are easily maintained.	—	—	—	—	—	—
Planned maintenance budget supports facility.	—	—	—	—	—	—

NAVAL FACILITIES ENGINEERING COMMAND
CUSTOMER QUALITY SURVEY

Maintenance of your Facility continues: — — — — —   —

Comments: _____

_____

_____

Section IX: Coordination and Communication with NAVFAC

The statements below refer to different aspects of how much or how well your command dealt with NAVFACENGCOM during the planning, design and construction phases of your facility. Please indicate how strongly you agree or disagree with these statements by checking the corresponding blank.

	Strongly Agree				Strongly Disagree	No Opinion/ Don't Know/
	5	4	3	2	1	Doesn't Apply
NAVFAC handled planning process well.	—	—	—	—	—	—
NAVFAC handled design process well.	—	—	—	—	—	—
NAVFAC handled construction process well.	—	—	—	—	—	—
NAVFAC handled maintenance turnover process well.	—	—	—	—	—	—
NAVFAC used our input during design process.	—	—	—	—	—	—
ROICC was responsive to our concerns during construction process.	—	—	—	—	—	—
Partnering during construction was a useful experience.	—	—	—	—	—	—
Acquisition strategy decision should have involved customer more.	—	—	—	—	—	—

Coordination and Communications comments:

_____

_____

_____

How satisfied are you with the **quality** of this facility?

Highly Satisfied	Somewhat Satisfied	Neutral	Somewhat Dissatisfied	Highly Dissatisfied	No Opinion
___	___	___	___	___	___

What are the **BEST** things about this facility?

1.

2.

3.

4.

5.

What are the **WORST** things about this facility?

1.

2.

3.

4.

5.

How satisfied are you with the NAVFAC
facility delivery process?

___	___	___	___	___	___
Highly Somewhat		Somewhat	Highly		No
Satisfied Satisfied	Neutral	Dissatisfied	Dissatisfied		Opinion

What was **BEST** about the process?

1.

2.

3.

4.

5.

What was **WORST** about the process?

1.

2.

3.

4.

5.

If you could influence NAVFAC to strengthen, change, or modify an existing service or offer new service;
what would you propose?

Thank you for your time and your thoughtful responses.

Focus Group Questions for Team Leader

Allow an hour - hour &1/2 for focus group discussion. The purpose of the discussion is to
determine consensus areas among the participants of their strongly held positive and/or negative
views.

The focus group questions relate to the questionnaire and are as follows:

1. How well does this facility support your mission? In other words, in what ways does this
facility help or hinder your?

2. Tell us about the environment inside the facility (air quality, trash, clean-up, sun/shade,
lighting, etc.)

3. How comfortable is this facility to be in? How about handicapped access?

4. How safe do you feel in or around this facility?

5. Tell us about the appearance of the facility. Looks good/not so good/what?

6. How do you feel about the planning that went into this facility? Were you involved?

7. How well is this facility maintained and how easy is it to do?

8. Coordination and communication are often mentioned as important parts of the planning, designing and construction of a new facility. How do you feel about this aspect of the Naval Facility Engineering Command's involvement in the delivery of this facility?

9. Overall, how satisfied are you with the quality of this facility?

10. Is there any NAVFAC product or service that you feel strongly contributed to the success or failure of your project.

11. If you could add, subtract, or change a service, what would you propose?

When the consensus held view shows a need for further evaluation and action ensure the problem is clearly defined. If required for definition, use "rough" sketch or photograph(s). Your final report should clearly communicate all positive/negative consensus areas.

The activity Point-of-Contact shown on page one of your report will receive feedback from the survey **following** its submittal and evaluation. The feedback will occur through the Engineering Field Division or Engineering Field Activity who administered the design and construction.

EXAMPLE OF GENERAL SERVICES ADMINISTRATION CRITERIA-BASED POE

To:	*Building Occupants and Users*
Due Date:	*May 1ˢᵗ, 1997*
Return to:	*U.S. Marshal Service (c/o Jim Falkenstrom)*
Re:	*Post-Occupancy Evaluation*
	Bruce R. Thompson U.S. Courthouse/Federal Building, Reno, Nevada

The General Services Administration is conducting a **Post-Occupancy Evaluation (POE)** of the Bruce R. Thompson U.S. Courthouse/Federal Building, Reno, Nevada. The purpose of this evaluation is to assess how well the building performs for those who occupy it in terms of the functions for which it was designed. The results of the study will be used to help improve the design of Federal Buildings and U.S. Courthouse facilities in the future.

Please respond only to those questions of the following survey that are applicable to you. Please indicate your answers by marking the appropriate blanks with an "X".

1. In an average work week, how many hours do you spend in the following types of spaces:

Hours	Court room(s)	Judicial Chamber Areas	Other Offices	Other Areas (specify)
0-5	☐	☐	☐	☐
6-10	☐	☐	☐	☐
11-15	☐	☐	☐	☐
16-20	☐	☐	☐	☐
21-25	☐	☐	☐	☐
26-30	☐	☐	☐	☐
31-35	☐	☐	☐	☐
36-40	☐	☐	☐	☐
More than 40 ☐				

Key for the following questions:	P = Poor (Much worse than expected)
	F = Fair (Somewhat worse than expected)
	G = Good (As Expected)
	VG = Very Good (Somewhat better than expected)
	EX = Excellent (Much better than expected)

2. Please rate the overall quality of the following areas of the building:

		P	F	G	VG	EX
a)	Courtroom	☐	☐	☐	☐	☐
b)	Judge's personal office (chambers)	☐	☐	☐	☐	☐
c)	Staff offices of the judge	☐	☐	☐	☐	☐
d)	Public corridors	☐	☐	☐	☐	☐
e)	Public restrooms	☐	☐	☐	☐	☐
f)	Storage	☐	☐	☐	☐	☐
g)	Elevators	☐	☐	☐	☐	☐
h)	Parking	☐	☐	☐	☐	☐
i)	Other (specify) _____	☐	☐	☐	☐	☐

Comments _____

3. Please rate the quality of the courtroom in terms of the following:

                                                        P   **F**   **G**   **VG**   **EX**

a) Adequacy of space .... ☐ ☐ ☐ ☐ ☐
b) Ceiling height.. ☐ ☐ ☐ ☐ ☐
c) Storage space .... ☐ ☐ ☐ ☐ ☐
d) Lighting... ☐ ☐ ☐ ☐ ☐
e) Acoustics... ☐ ☐ ☐ ☐ ☐
f) Temperature/Temperature Controls .... ☐ ☐ ☐ ☐ ☐
g) Image/Aesthetics.. ☐ ☐ ☐ ☐ ☐
h) Security/Security Systems.. ☐ ☐ ☐ ☐ ☐
i) Flexibility of Use .... ☐ ☐ ☐ ☐ ☐
j) View to the Outside .... ☐ ☐ ☐ ☐ ☐
k) Other, specify _____ .... ☐ ☐ ☐ ☐ ☐

Comments _____

_____

4. Please rate the quality of the Chambers and/or judicial staff office areas in terms of the following:

                                                        P   **F**   **G**   **VG**   **EX**

a) Adequacy of space .... ☐ ☐ ☐ ☐ ☐
b) Storage space .... ☐ ☐ ☐ ☐ ☐
c) Lighting... ☐ ☐ ☐ ☐ ☐
d) Acoustics... ☐ ☐ ☐ ☐ ☐
e) Temperature/Temperature Controls .... ☐ ☐ ☐ ☐ ☐
f) Image/Aesthetics.. ☐ ☐ ☐ ☐ ☐
g) Security/Security Systems.. ☐ ☐ ☐ ☐ ☐
h) Flexibility of Use .... ☐ ☐ ☐ ☐ ☐
i) View to the Outside .... ☐ ☐ ☐ ☐ ☐
j) Other, specify _____ .... ☐ ☐ ☐ ☐ ☐

Comments _____

_____

5. Please rate the quality of public areas in the building in terms of the following:

                                                        P   **F**   **G**   **VG**   **EX**

a) Adequacy of space .... ☐ ☐ ☐ ☐ ☐
b) Ceiling height.. ☐ ☐ ☐ ☐ ☐
c) Storage space .... ☐ ☐ ☐ ☐ ☐
d) Lighting... ☐ ☐ ☐ ☐ ☐
e) Acoustics... ☐ ☐ ☐ ☐ ☐
f) Temperature/Temperature Controls .... ☐ ☐ ☐ ☐ ☐
g) Image/Aesthetics.. ☐ ☐ ☐ ☐ ☐
h) Security/Security Systems.. ☐ ☐ ☐ ☐ ☐
i) Flexibility of Use .... ☐ ☐ ☐ ☐ ☐
j) View to the Outside .... ☐ ☐ ☐ ☐ ☐
k) Other, specify _____ .... ☐ ☐ ☐ ☐ ☐

Comments _____

_____

6. Please rate the quality of materials used in this building:

	P	F	G	VG	EX

Courtrooms
a) Floors .... ☐ ☐ ☐ ☐ ☐
b) Ceilings .... ☐ ☐ ☐ ☐ ☐
c) Walls .... ☐ ☐ ☐ ☐ ☐

Chambers/Judicial Staff Areas
a) Floors .... ☐ ☐ ☐ ☐ ☐
b) Ceilings .... ☐ ☐ ☐ ☐ ☐
c) Walls .... ☐ ☐ ☐ ☐ ☐

Other Offices
a) Floors .... ☐ ☐ ☐ ☐ ☐
b) Ceilings .... ☐ ☐ ☐ ☐ ☐
c) Walls .... ☐ ☐ ☐ ☐ ☐

Public Areas
a) Floors .... ☐ ☐ ☐ ☐ ☐
b) Ceilings .... ☐ ☐ ☐ ☐ ☐
c) Walls .... ☐ ☐ ☐ ☐ ☐

7. Please rate the quality of the building as a whole in terms of the following:

	P	F	G	VG	EX

a) Aesthetic quality/"look" of the exterior .... ☐ ☐ ☐ ☐ ☐
b) Aesthetic quality/"look" of the interior .... ☐ ☐ ☐ ☐ ☐
c) Ability of visitors to find their way around .... ☐ ☐ ☐ ☐ ☐
d) Amount of space .... ☐ ☐ ☐ ☐ ☐
e) Environment in the building (lighting, acoustics
temperature, temperature control, etc.) .... ☐ ☐ ☐ ☐ ☐
f) Functional layouts/flows .... ☐ ☐ ☐ ☐ ☐
g) Availability of conference rooms .... ☐ ☐ ☐ ☐ ☐
h) Adaptability to changing uses .... ☐ ☐ ☐ ☐ ☐
i) Security (during business hours) .... ☐ ☐ ☐ ☐ ☐
j) Security (after business hours) .... ☐ ☐ ☐ ☐ ☐
k) Housekeeping and Maintenance .... ☐ ☐ ☐ ☐ ☐
l) Provision of natural light and views .... ☐ ☐ ☐ ☐ ☐
m) Other, specify _____ .... ☐ ☐ ☐ ☐ ☐

Comments _____

_____

8. Please select and rank in order of importance the five qualities of the building (listed in Question 7) that matter most in making a productive work environment for you:

1. _____
2. _____
3. _____
4. _____
5. _____

9. Please identify important spaces and facilities that are lacking in the building:

10. Do you work primarily in (please check one):

_____ An open office are or special rooms (computer or file rooms)?

_____ A cubicle workstation with panels surrounding

_____ A private office

_____ Other, please describe _____

11. Please rate the quality of your primary work space in terms of the following:

	P	F	G	VG	EX
a) Size	☐	☐	☐	☐	☐
b) Configuration	☐	☐	☐	☐	☐
c) Storage	☐	☐	☐	☐	☐
d) Noise level	☐	☐	☐	☐	☐
e) Number and location of electrical outlets	☐	☐	☐	☐	☐
f) Number and location of data/telecom outlets	☐	☐	☐	☐	☐
g) Lighting Level	☐	☐	☐	☐	☐
h) Freedom from glare	☐	☐	☐	☐	☐
i) Temperature/Temperature control	☐	☐	☐	☐	☐
j) Air Quality (stuffiness, odors, etc.)	☐	☐	☐	☐	☐
Computer users only					
k) Keyboard location	☐	☐	☐	☐	☐
l) Glare on screen	☐	☐	☐	☐	☐
m) Number and location of data/telecom outlets	☐	☐	☐	☐	☐

12. Thinking about the entire building and your workplace, what two things do you like the most?

1. _____

2. _____

13. What do you like least about the building and your workplace?

1. _____

2. _____

14. Comparing this building to your previous work environment, which do you prefer?
_____ Please explain why:

Your Department: _____ Located on Floor:_____

PLEASE INDICATE ON THE ATTACHED FLOOR PLANS THE LOCATION OF YOUR PRIMARY
WORKSPACE.

Thank you for your assistance.

UNITED STATES
POSTAL SERVICE

Post Occupancy Evaluation

Instructions

Date: [_____]

To: Postmaster/Facility Manager
Manager, Administrative Services Office

To better serve you, Facilities conducts a Post Occupancy Evaluation program to determine how well a recently completed facility serves your needs and to identify areas for improvement in future facilities. Your participation in this survey is valued and important since both positive and negative comments will help improve the facilities.

The Post Occupancy Evaluation Questionnaire is attached. It would be appreciated if you would complete and answer all questions to the best of your ability. You may wish to ask your staff for input on those issues with which they are more familiar.

- One copy of the questionnaire is to be completed by the Postmaster/Facility Manager.

- Another copy of the questionnaire is to be completed by the Manager, Administrative Services Office.

- The questionnaire is to be completed between four and six months after the facility is occupied.

Most questions are statements which ask if you "strongly agree" (5) or "strongly disagree" (1). You may select any of the numbers between 1 and 5 to indicate the extent of your agreement. You may comment on any question on the form or attach more detailed comments. Please provide comments on any item where you check "strongly disagree". If a question does not apply to your facility (for example, if the questions asks about parcel lockers or carrier platforms and your facility does not have them) indicate NA for not applicable.

Your responses will be reviewed by the Manager, Design & Construction at the FSO responsible for construction of the building, and will be forwarded to Headquarters. As part of this program, some facilities will be selected for a site visit and more in-depth study. Findings on all facilities will be summarized to identify patterns and to improve USPS standard design criteria and procedures on future postal facilities.

Please return the completed questionnaire to me by [_____]. If you have any questions please call me at [___-___-___]. Thank you for your assistance.

Manager, Design & Construction
Facilities Service Office

UNITED STATES
POSTAL SERVICE

Post Occupancy Evaluation Questionnaire

Facility name:

Occupancy date: Size: < 9,000 nsf [ ]   > 9,000 nsf [ ]

Manager name: Title:

City: State:

Telephone: Date:

Process Management Indicators   yes   no

A1   Does this new building provide operational space to meet your 10-year needs?   [ ]   [ ]

A2   Were you furnished with original project milestone dates and
      and kept informed of changes to these dates?   [ ]   [ ]
      (Dates listed in the Investment Cost Sheet and the Decision Analysis Report.)

Please answer the questions by filling in the boxes with a number on a scale of 1 to 5 as follows:

Strongly agree	5
Agree	4
Neither agree nor disagree	3
Disagree	2
Strongly disagree	1
Not Applicable	NA

1   The Inspection Service attended sufficient design review meetings.   [ ]

2   The site and building design allow adequate space for future expansion.   [ ]

3   The customer parking is easy to enter and exit.   [ ]

4   Traffic flow through the customer parking area is good.   [ ]

5   The number of customer parking spaces is sufficient.   [ ]

6   The number of employee parking spaces is sufficient.   [ ]

7   Drainage is sufficient to avoid ponding.   [ ]

8   Exterior signage and pavement markings are sufficient and well located.   [ ]

9   Landscaping is appropriate; attractive, does not create security problems.   [ ]

10   Building identity and signage is visible from the customers' approach.   [ ]

11   The roof system performs well and does not leak.   [ ]

12   Loading dock access, stairs and ramps are convenient.   [ ]

13   The number and type of loading dock leveling devices is appropriate.   [ ]

14   The covered carrier loading slab is adequate in location and size.   [ ]

15   Platform doors are adequate, operate well and provide security.   [ ]

16	The self service area is effective in attracting customers.	☐
17	The type and number of self service equipment is appropriate.	☐
18	Parcel lockers are in an appropriate location, size and quantity.	☐
19	The number of service counters appears sufficient for the next ten years.	☐
20	Electronic article surveillance and CCTV systems are effective.	☐
21	Workroom flooring material is appropriate; durable and comfortable.	☐
22	Storage space is sufficient.	☐
23	Workroom lighting is appropriate; sufficient, well distributed with minimal glare.	☐
24	Office space is appropriate in size, quantity and arrangement.	☐
25	The heating system provides comfortable conditions.	☐
26	The cooling system provides comfortable conditions.	☐
27	The telephone system, including jack locations, is appropriate.	☐
28	Building construction was completed without extensive delays.	☐
29	Punchlist items were completed in a timely manner.	☐
30	The contractor trained the staff sufficiently to operate the building systems.	☐
31	Sufficient training was provided to operate the EAS, IDS and CCTV systems.	☐
32	A set of "as-built" drawings was received in a timely manner.	☐
33	Operating and maintenance manuals are well arranged and easily understood.	☐
34	Contractor support during warranty period was sufficient.	☐

Please identify features that should be applied to future projects,
problems that should be corrected at future projects and other comments:

_____

_____

_____

_____

_____

_____

_____

_____

Appendix E

Supplemental Information to Chapter 6

COMPANIES OFFERING ONLINE SURVEYS AND/ OR POLLING SERVICES[1]

In general, each package will have the following features, to some level of ease:

- installation and integration of software into a system
- supporting documentation, such as online help, tutorial, or a guide on how to make a survey
- the ability to add, edit, and manage templates provided
- options for building single- and multiple-page forms and branching to other questions within or between pages
- ability to scan e-mail or data files as the results come in
- file management features such as importing and exporting data, data cleaning, and record keeping
- the ability to post surveys on the Web and provide support to a server
- data analysis tools and types of analysis available
- options to chart and present data
- overall ease of using the product and its user interface

THE CHANGING CONTEXT OF ONLINE COMMUNICATIONS

Today, anyone with a cyber address is inundated with unsolicited messages and unnecessary communications, often originating from within their own organizations. The exponential growth of junk e-mail in

_____

[1]See King (2000) for reviews of the software.

TABLE E-1 Companies Offering On-line Surveys or Polling Services.

EZSurvey 2000 www.raosoft.com	SurveySolutions for the Web 3.0 www.perseus.com
SurveyCrafter Professional 2.7 (previously MarketSight 2.5) www.surveycrafter.com	WebSurveyor www.websurveyor.com
Survey Select 2.1 and Survey Select Expert 4.0 www.surveyselect.com	Zoomerang www.zoomerang.com

recent years is a phenomenon termed "spam" (noxious, unwanted e-mails). Using current communications technology, a single cyber-marketing company can send half a billion personalized ad mails via the web every day. It is estimated that it costs Internet users worldwide $US 9-billion ($CDN 14-billion) annually to receive junk e-mails (Hargreaves, 2001). In this environment, people may not bother to open unsolicited e-mail or to agree that a survey be sent to them.

The low response rate for online surveys might also reflect a general mistrust of electronic communication. For example, unbeknownst to users, their consumer information may be gleaned while they visit Web sites. Then this information can be sold for large sums of money and so it escalates. Having been damaged by tempting messages, such as the "I love you" virus, computer users may now be more cautious of electronic invitations, limiting their willingness to participate in online surveys. This would apply to wide-cast cybersurveys, and less so to e-surveys sent through an organization's proprietary network/intranet.

There are Web users who pay for their time online. That could deter some from spending valuable minutes to fill out a survey. Eliminating these potential respondents both lowers the response rate and might also add a bias based on income.

On the other hand, the cost to connect is steadily coming down and there are increasing opportunities for the general public to access the Internet. Businesses such as easyEverything <www.easyeverything.com>, Kinko's <www.kinkos.com>, and Get2net <www.get2net.com> are filling storefronts in city centers. At easyEverything in Manhattan there are 800 terminals with Internet access and one dollar ($US1) buys two hours of connectivity. According to Pike (2001), there is an interesting cast of characters accessing the net at 11PM on Saturday night at the Times Square location. Kinko's offers a fast connection to surf the Internet and use Microsoft's complete Office Suite (Word, Excel, PowerPoint) for thirty cents a minute. Get2net has free Internet kiosks at select locations, however keyboards are awkward, access slow, and there's lots of advertising.

DETAILS OF WHO IS ONLINE AND WHERE THEY ARE GEOGRAPHICALLY

The number of people accessing the Internet continues to increase at a phenomenal rate. In 1995 The Internet Society estimated that between 20 to 40 million people around the world had access to the Internet. Nua Internet Surveys (Nua, 2001) estimated that number to have grown to 201 million worldwide in 1999, and up to 407 million by 2000. See Table E-2.

Early Internet users (circa 1995) tended to be young, white males with high socioeconomic status. Recent studies suggest that as more people use the Internet and World Wide Web, there is a demographic shift and that Internet users are beginning to represent more of the general population. More households have Internet connections. The US Department of Commerce (1999) reported that the number of households connected to the Internet increased from 18.6% in 1997, to 26.2% in 1998.

The take-up of electronic communications is faster than any other "disruptive technology" of the 20th century—namely electricity, the telephone, and the car. In general, a medium is considered a "mass medium" when a critical mass of people (about 16% of the population, or 50 million for the USA) has adopted the inno-

TABLE E-2 Top Ten Countries with Internet Users - Number and Percentage of Users.

Country	Population (in million)	Internet Users (in million)	% of Population on Internet
Australia	19	7.4	38.9%
United States	276	91.0	33.0%
Canada	31	9.7	31.3%
Japan	127	29.0	22.8%
Germany	83	18.9	22.8%
United Kingdom	60	18.8	31.3%
South Korea	46	14.0	30.4%
France	59	10.7	18.1%
Italy	58	6.6	11.4%
China	1,300	10.0	0.8%

Source: Morgan Stanley Dean Witter, 2001.

vation (Markus, 1990). It took 38 years for radio to reach this level of adoption. Television took 13 years and cable television reached a critical mass in 10 years. Depending on the various estimates on the number of Internet users, the medium has already reached critical mass or will certainly be there by 2002, just 8 years after its emergence as a consumer medium (Neufeld, 1997).

Although a large number of people access the Web, in 1998 they accounted for less than one third of the overall USA population (Kaye and Johnson, 1999). Estimates vary, and as much as half the USA population may be connected. The fast take-up of this medium is rapidly changing the profile of who's online, making less relevant some of the lessons-learned and sampling issues from earlier work. The trends suggest that the number of users will continue to grow, will better reflect the overall population, and that upwards of 80% of Internet users will access the system daily. Such a user base would provide a reliable population from which to sample and generalize findings.

REFERENCES

Hargreaves, D. 2001. Junk e-mail costs online surfers $14-billion a year: EU report. *Financial Post*. February 3. p. D9.

Kaye, B. and T. Johnson. 1999. Research Methodology: Taming the Cyber Frontier. *Social Science Computer Review*. Volume 17, No 3, pp. 323-337.

King, N. 2000. What are they thinking? *PC Magazine*. February 8, pp. 163-178.

Markus, L. 1990. Toward a "critical mass" theory of interactive media. In Fulk, J. and C. Steinfield (eds.) *Organizations and communication technology*. pp. 194-218. Newbury Park, CA: Sage.

Morgan Stanley Dean Witter. 2001. In *Infoworld*. March 12, p. 16.

Neufeld, E. 1997. Where are the audiences going? *MediaWeek*. May 5, pp. S22-S29.

Nua Internet Surveys. 2001. http://www.nua.ie/surveys/how_many_online

US Department of Commerce. 1999. Falling through the net: defining the digital divide. *National Telecommunications and Information Administration*. 27 pages.

Appendix F

Chapter 5 from *Post-Occupancy Evaluation Practices in the Building Process: Opportunities for Improvement,* National Academy Press, 1987

TRENDS, CONCLUSIONS, AND RECOMMEDATIONS

Based on observed trends, the committee makes the following conclusions and recommendations to be considered by government agencies and private organizations responsible for construction programs in general and for conducting POEs in particular. These recommendations recognize the current lack of institutional support for this field, as well as the need to generate a reliable and comprehensive data base. To that end, the committee proposes measures that will: (1) make POE a more systematic process with rigorous procedures, (2) lay the groundwork for a data base of knowledge on building use and performance, and (3) establish a clearinghouse to assemble, maintain and disseminate information generated by POEs.

The committee presents trends, conclusions, recommendations and discussions (in that order) in three sections: (1) policy-related topics that focus on broad policies that should be instituted to make POE more useful and widely used, (2) building process-related topics that focus on procedures in the uses of POE, and (3) technology and techniques-related topics that identify innovative ways in which POEs can be improved.

ITEMS RELATED TO POLICY IN THE BUILDING PROCESS

This section presents three policy-related topics: (1) monitoring building quality and performance, (2) POE data base and clearinghouse information, and (3) POE data and litigation. The committee recommends courses of action that can be implemented to make POE more beneficial to government agencies and private organizations, as well as to improve the planning, design, construction and operation of facilities.

Monitoring Building Quality and Performance

Trend 1. Quality assurance programs are used by many manufacturing concerns to raise consumer confidence and to compete more effectively in world markets. POE addresses a significant part of quality assurance in the building industry. As each facility is evaluated in use, the existing quality of materials and design concepts can be critically assessed, and design criteria can be changed to produce better facilities in the future.

Hidden high operating costs, costly repairs, and dysfunctional facilities have made administrators more aware of the need for quality buildings. This is especially true for those institutions that build many facilities on a recurring basis.

POE also constitutes an "auditing" tool that can be used by the knowledgeable client. Together with construction audits and other recognized financial accounting practices, POE can track performance of a project, document costly changes to the program requirements, and identify critical strengths or weaknesses associated with a particular facility type. POE can be used as a parallel track in the design and construction process, tracking decisions, changes, and outcomes.

Conclusion 1. More emphasis is being placed on quality in our society today. Occupants of facilities expect that same quality in terms of building performance. Organizations in the private and public sectors are concerned about the price/performance relationship for new facilities and, therefore, want to develop responsive buildings for the lowest possible cost.

Recommendation 1. Government agencies manage and operate a significant real estate portfolio for their own account. This includes offices, hospitals, housing and special use facilities. Agencies should be learning and benefiting more from their extensive design, construction and operational experience. Through POE, they should be applying the lessons learned to reduce operating costs, to design environments that improve productivity, and to build facilities that respond to the rapidly changing requirements of institutional users and clients.

Discussion 1. By evaluating the performance of new and existing facilities in terms of how well they work for the user, agencies can make trade-offs on future projects and target features that have the greatest return in assuring building quality and performance. Furthermore, professionals in the design, construction and facilities management community can exchange information with one another through associations, conferences and written presentations. By exchanging information generated by POEs about buildings in use, they will greatly expand knowledge about how to achieve better quality buildings.

Development of Data Bases and Clearinghouses

Trend 2. Today there is the technology and capability to develop electronic data bases for use by various participants in the building industry. Some large corporations have already begun developing these data bases to disseminate information to subscribers and other interested users. Certain industries have established clearinghouses to collect, organize, archive and disseminate specialized information. Clearinghouses have helped to advance practice in those fields that build upon precedent and other professionals' work. In the building industry, several professional groups (such as the International Facilities Management Association and the Building Owners and Managers Association Exchange) are involved in establishing clearinghouse activities. At this time, however, they are not designed to handle POE information.

Conclusion 2. Because the operation of facilities is becoming increasingly complex, the sharing of knowl-

edge and experience takes on added importance. POE results can be organized into a data base format and can be made available to subscribers through a clearinghouse, subscription service, or an electronic data base. Standardized documentation and data collection can be used, and government building projects can be entered and evaluated as part of the data base. The clearinghouse would manage the information and provide a central source of expertise.

Recommendation 2. Government agencies, together with private sector organizations, should create and support: (1) an on-line data base that would contain POE results, design criteria, and other design guidelines, (2) a clearinghouse consisting of electronic data bases, and case studies, and (3) POE networks, directories, conferences, and other ways within government agencies to expedite the exchange of knowledge.

Discussion 2. Various academic and research-oriented associations already have extensive, informal POE networks. Currently, these networks are developed and maintained through voluntary efforts. Federal agencies, at very little expense, could put together a directory of individuals and groups who conduct POEs or who are interested in doing so. Various forms of information exchanges and other supporting materials could be assembled. The data base would be kept updated by incoming POE reports, by special studies related to important design concepts, and by other research. It would be set up so as to be accessible through national networks already established.

A centralized capability to organize, collect and disseminate POE-based knowledge is critically needed. New building projects are usually begun without knowledge of how previous solutions have fared; too often new design is based on architectural trends, aesthetics or first-cost considerations. The cost to the government is enormous as novel building designs fail, buildings do not satisfy the needs of users, and cost overruns mount because of unanticipated problems.

A clearinghouse, as envisioned by the committee, would primarily use electronic data bases that could be accessed by private and public sector users on a fee-for-service basis.

POE Data and Litigation

Trend 3. Increased use of litigation in our society raises a concern about the possible use of POE data in lawsuits. The fear is that responsibility will be attributed from POE data to certain parties, and lawsuits and costly legal expenses may result.

Conclusion 3. The possibility of POE data used in lawsuits may potentially have a crippling effect on the continued development of the field. Actions should be taken to safeguard the use of POE results.

Recommendation 3. In conjunction with other policy actions, adequate information controls and safeguards should to be developed and implemented in any POE program. A legal and ethical code is also required to cover POE use. Public sector agencies, working through the Federal Construction Council, should request the Building Research Board (or similar organization) to develop appropriate procedures and safeguards.

Discussion 3. Safeguards need to be built into any POE program to insure accurate reporting, to minimize nuisance suits, and to protect the parties involved in the design and construction of buildings. If POE becomes associated with punitive litigation in this way, agency personnel may refuse to do POEs. The results of POEs should remain confidential until clearance by the client organization.

ITEMS RELATED TO PROCEDURES IN THE BUILDING PROCESS

This section considers four topics related to POE practice and procedures: (1) the building performance concept and standards, (2) changing human requirements and building technologies, (3) user participation and training, and (4) economics. The committee believes that POE can significantly improve buildings by promoting research-based programming and design. The results of POE can be used to identify key factors about building performance that make the operation and management of buildings more efficient and cost effective.

The Building Performance Concept and Standards

Trend 4. Higher quality buildings can be developed as POE data bases come into general use. The results of POE will provide designers with an empirical base on the performance of buildings that can be used to assess other buildings and to evaluate new design concepts.

The results, factored into the building process through updated codes and revised criteria, will promote greater quality design solutions. Unified and accepted standards allow for the communication and comparison of data from individual studies. Such standards develop a higher level of professionalism in the field; practitioners adhere to these practices, thus allowing for the comparison of findings and the interpretation of results.

Conclusion 4. Buildings are designed based on certain goals and performance requirements that are further clarified by defining explicit, often quantitative, performance criteria and by establishing a range of measured values that will satisfy those criteria. Since many POEs in current practice are ad hoc in character, there is little basis for comparisons or for valid inferences to be drawn. A systematic POE program would enhance the ability of a regulatory agency to verify compliance with the performance features of its codes and standards. Rigorous POEs would fill this need, increasing design flexibility as well.

Recommendation 4. Key indicators and reliable, objective building performance measures should be developed for use in POEs, and as a basis for design criteria, as well as standards and guidelines for a variety of common facility types. In addition, the description and documentation of the buildings being evaluated should be improved.

Discussion 4. A set of key indicators, similar to the economic indicators used in evaluations of the state of the economy, should be used in POEs. These key indicators, associated with other, more standardized methods and procedures, would lead to more reliable results that could be more easily communicated. In addition, efforts should be made to develop performance requirements (e.g., purpose, description, assessment, conclusions of lessons learned, strengths and weaknesses) that can be used at different levels of evaluation—from walk-through POE ratings to more in-depth diagnostic POEs.

Several levels of POE investigation can be undertaken. Each of these levels has somewhat different objectives and requires a somewhat different set of methodologies, procedures, and related formats. These procedures should be standardized so as to allow for the comparability of information and results. Such standardization, including how to handle exceptions, would allow data to be entered into a data base and could be made available to all pertinent government agencies.

A taxonomy of buildings that describes meaningful categories of features, materials and systems can be developed to create a common basis for comparisons and evaluations. The physical environment that is being evaluated in a POE needs to be adequately described so that one study of that facility type can be related to another. A set of descriptor categories as well as a set of physical, objective measures (e.g., lighting levels on the work surface) should be included.

Changing Human and Building Parameters

Trend 5. Users expect environments that are responsive to their needs. Occupants are the critical element in helping organizations to achieve their mission, and facilities must support the needs of building occupants. The introduction of new materials and technologies into the building industry will continue as producers develop new products and applications. This will present opportunities for new design strategies and solutions; it will also present new dangers. Some of these products are tested and evaluated in a laboratory setting, but problems are only identified after an extended period of use (e.g., gases being given off from some office products contributing to poor indoor air quality) or when a disaster occurs (e.g., toxic fumes being produced by the burning of certain plastic materials in furniture). There is a need for evaluating these materials and technologies in use when they are combined with other products or put to unusual and novel uses.

Conclusion 5. New technologies, a consumer ethic, and more education are changing the way people use designed environments. Environments must become more flexible to accommodate frequent changes, and they must be more responsive to provide for newly emerging needs. New materials and technologies allow the designer to create specialized spaces in buildings to house a variety of activities.

New materials (e.g., plastics, bonding agents, and sealants) are being developed for use in buildings, building systems and furniture. Similarly, new technologies are being introduced into buildings that allow for improved building operation or new design options. New computer-based technologies that augment the activities of building occupants are placing new demands on building systems and performance.

Recommendation 5. POE programs should be developed that allow facility management to assess and plan for the changing requirements of building occupants. POEs can be used to evaluate existing buildings, regularly assess users' perceptions of the facilities, and plan for necessary changes based on user needs.

POE practices should be employed to evaluate new materials and technologies in actual use. Prototypes or representative cases should be identified and studied; the results from evaluations could then be generalized to future applications.

Discussion 5. Organizations change to meet new conditions; as a result, individuals within organizations frequently move or change activities. A POE program that regularly evaluates facilities from the users' perspective can respond to these ongoing changes. New needs can be anticipated; facilities can be fine-tuned or retrofitted. Building maintenance priorities can be established on safety concerns and occupants' needs. Facility management can then supply environments that are flexible, respond to the changing needs of users, and are satisfying places in which to live and work.

A POE program, especially one that incorporates results into clearinghouses and electronic data bases, can be used to spot potential problems or trends across various facility types before a disaster occurs or a serious health hazard develops. With an early identification program, agencies can avoid possible litigation or costly retrofits to resolve health-related problems. Liaisons with other agencies that act on behalf of the public welfare or conduct research related to health and safety problems could be established. Liaisons with product testing laboratories, manufacturing associations or professional societies would also promote the sharing of results and rapid communication if potential problems were identified. POE programs could provide a testing capability for building products, materials and technologies in actual use. These evaluations could also furnish valuable research data for product modifications.

User Participation and Training

Trend 6. *User participation:* Focus groups, user panels, surveys and other forms of market research are used extensively in consumer product development to establish user preferences and product acceptance. User behavior patterns and human factor considerations have been recently added to product development efforts.

Training: Organizations that develop and manage facilities for their own users are finding it increasingly necessary to professionalize their in-house staff, which is expected to be proactive (i.e., anticipate change and plan for change). Such an expectation of professionalism requires more training of existing in-house staff.

Conclusion 6. *User participation:* User participation is an accepted practice, which in the building field can boost morale, improve office productivity, and provide other user benefits.

Training: As organizations professionalize their facilities management staff, they need to provide specialized training and education programs. These programs should deal with planning and implementing change, using the environment to support organizational objectives, involving users in the planning process, and implementing facilities management programs.

Recommendation 6. *User participation:* POE programs that solicit end user feedback and information should be used to heighten participation in the design of new facilities or in improving existing ones.

Training: Certification or other training programs should be developed to educate agency personnel or their consultants regarding concepts and techniques of conducting POEs.

Discussion 6. *User participation:* Typically, client representatives of the building owner make decisions for most people in the organization. Often, they do not have first-hand experience of various functions, nor do they know the personal preferences of individuals. It would be more effective to have end users participate by expressing attitudes, personal preferences, behavioral styles, and other characteristics of a more personal nature. This higher level of participation by users would provide a richer representation of user needs from which to develop new design solutions.

Training: Pilot training programs could be created by knowledgeable POE practitioners to train in-house personnel or consultants on how to do POEs. This could be done through universities and technical schools. Printed instructional material could be supplemented by videotaped materials of case studies, or tutorials documenting how a POE is conducted. Seminars or workshops could also be established for designers, facility managers, building operators, and real estate consultants. Professional associations such as the Inter-national Facilities Management Association, the Building Owners Management Association, the National Office Products Association, and others intimately involved with the building industry should be enlisted in this training effort.

Economics

Trend 7. Life-cycle costing, including the costs of operation, maintenance, and other facility-related activities, is an increasingly important consideration for institutions that develop their own facilities. Facility managers can adopt POE procedures to project the quality of a facility that they build, as well as to evaluate its performance over time.

Conclusion 7. Increasingly, organizations are becoming more aware of the value of the fixed asset base which they own, manage and operate. Facilities management as a professional occupational category has also grown, as owners adopt a more active posture regarding the management of their facilities. It is expected that POE will generate significant cost savings by improving design criteria, by correcting problems that are discovered after building occupancy, and by improving the overall building stock over the long term.

Recommendation 7. POEs should become part of the management process used by facility managers, building operators, and others responsible for fixed asset management, new facility development, or design. Research on building economics and the overall life-cycle costs associated with a facility should be conducted. Research should also be done on the costs of POEs, including savings that are realized as a result of POEs.

Discussion 7. There has been a growing recognition of the economic importance of high quality buildings and their management as fixed assets. Similar to the human resource or information resources management functions, facility management is concerned with the day-to-day operations and, occasionally, with new building projects.

The economics of building occupancy are related to the housing of personnel, technology and various functions of an agency and should be viewed from a building life-cycle perspective. Similarly, POE should be seen as part of the overall building process, and not as singular case studies. To that end, more economic research is needed to document savings and opportunities.

As POE is increasingly used to provide the industry with empirical data about buildings in use, POE results can document manufacturers' claims, give performance profiles of individual buildings systems, and provide information about possible trade-offs.

ITEMS RELATED TO INNOVATIVE TECHNOLOGIES AND TECHNIQUES

This section reviews four areas of POE practice pertaining to: (1) smart buildings, (2) computer-based systems, (3) simulations, and (4) mathematical modeling.

Smart Buildings

Trend 8. It is now possible to build into a facility the capability to monitor constantly or frequently occupants' responses. Individuals are able to provide feedback to facility managers through interactive, computerized "check-out" procedures. It may soon become possible to develop electronically monitored environments that respond to commands of building occupants. There will also be the capability of providing electronically adjusting environments that automatically change to meet the needs of the occupants.

Conclusion 8. With the introduction of electronic technology and building control systems into most facility types, new opportunities exist to develop more sophisticated evaluation methods and procedures.

Recommendation 8. The committee encourages the use of existing and new integrated building monitoring systems (such as security sensors, video monitors, and telephones) to provide data and to record occupants' feedback in response to building conditions. Research should be undertaken to develop on-line sensors, wear and tear indicators, and other potentially beneficial applications of monitoring technology to be used in POE programs.

Discussion 8. Many existing building monitoring systems and technologies routinely report on ambient conditions or on ongoing activities in buildings. Security cameras and movement sensors that regulate lighting, telephones, and other systems can be used to provide POE data and feedback to facility managers.

Computer-Based Systems

Trend 9. New computer-based technologies offer designers and end users the capability of visualizing and testing design concepts before they are actually built. They also offer more dynamic ways of sharing information and examining "what if" options in design.

Conclusion 9. How results of POE work or how graphical data are presented to an untrained audience is important. Long, written reports or complex graphics can lead to incorrect conclusions. Computer-based technologies, including computer-aided design (CAD) systems, offer many opportunities to educate an untrained audience and to communicate effectively new information.

Recommendation 9. The committee recommends the development of computer-based reporting formats that give POE practitioners the ability to communicate effectively their findings to a nontechnical audience. Alternative reporting and presentation formats should also be investigated.

Discussion 9. A CAD system could be linked to a POE data base to provide feedback on particular design configurations and strategies. CAD-generated materials could be reviewed and tested with prospective occupants before building construction begins. Psychological imaging, problem solving, and idea generation exercises could be used to augment the CAD capability in order to produce realistic images of actual occupancy experiences. For example, in Japan some developers already use CAD systems to help people design new homes. After prospective buyers play "what if" games with the sales person, a particular design is selected, and a CAD system produces information for the prefabrication of the housing units, including scheduling and delivery to the building site.

Simulations

Trend 10. Increasingly, electronic simulations are used to provide realistic experiences of actual live situations. The pilot trainer simulator is an example of this trend.

Simulators provide airplane pilots with the actual experiences, perceptual information, and other realistic inputs to simulate a situation or set of conditions that they might experience. Computers simulate instrumentation readings that appear in response to the pilot's control changes, and video displays provide them with realistic views of what they will encounter at different airports.

Conclusion 10. Computer simulations, used now in other industries, will provide POE practitioners with the tools and techniques to anticipate the findings of a POE before the building is built and occupied, enabling the end user to have input at the conceptual and design phases of the building process.

Recommendation 10. Computer simulations, such as full scale and smaller scale mock-ups, should be developed and used to complement POEs.

Discussion 10. Today, various psychological or other tests are used to evaluate situations such as job functions, living in a space capsule, or college performance. Some manufacturers have developed software programs that allow the evaluation of a machine part before it is produced. Simulation or evaluation tests could be developed to anticipate the significant findings that might be uncovered by POEs. While they will not remove the need for POEs, these simulations could provide critical feedback to designers.

Mathematical Modeling

Trend 11. Mathematical models are used to evaluate various economic, physical and political scenarios that could develop under certain conditions or assumptions. Scientists studying weather, geological events such as earthquakes, regional ecologies, and other natural systems rely on mathematical computer modeling to simulate possible outcomes and their likelihood of occurrence.

Conclusion 11. With availability of powerful computers and sophisticated software, it is becoming easier to use mathematical models for environmental design evaluations such as POEs.

Recommendation 11. Ways should be examined that allow the use of POE in early phases of building programming and design, before a building is actually completed and occupied. Simulations to replicate POEs, mathematical or statistical formulations, or expert computer systems may allow these types of pre-occupancy evaluations to be carried out.

Discussion 11. Models using POE information could be developed with the goal that at some time in the future the building industry will be able to do POE-type testing early in the programming phase of the building process. Planners would be able to ask "what if" questions and test them under various occupancy scenarios. Once POE findings are systematized, as called for in this report, it will become possible to apply mathematical and statistical models in the hope that some better predictability in design can be achieved.

Bibliography

Allen, T. (1977). *Managing the Flow of Technology.* Cambridge, Mass.: MIT Press.

American Society for Testing and Materials. (1987). *ASTM Standards on Computerized Systems.* West Conshohocken, Pa.

American Society for Testing and Materials. (1999). *ASTM Standards on Building Economics.* West Conshohocken, Pa.

American Society for Testing and Materials. (2000). *ASTM Standards on Whole Building Functionality and Serviceability.* West Conshohocken, Pa.

Amiel, M.S., and Vischer, J.C. (1997). *Space Design and Management for Place Making—Proceedings of the 28th Annual Conference of the Environmental Design Resarch Association.* Edmond, Okla.

Ang, G., et al. (2001). *A Systematic Approach to Define Client Expectation to Total Building Performance During the Pre-Design Stage.* Proceedings of the CIB 2001 Triennial Congress. Rotterdam, Holland.

Argyris, C. (1992). *On Organizational Learning.* Cambridge, Mass.: Blackwell, Business.

Argyris, C. (1992). Teaching smart people how to learn. In Argyris, C., (Ed.) *On Organizational Learning.* Cambridge, Mass.: Blackwell Business. pp. 84-100.

Argyris, C., and Schon, D. (1978). *Organizational Learning.* Reading, Mass.: Addison-Wesley.

Aronoff, S., and Kaplan, A. (1995). *Total Workplace Performance: Rethinking the Office Environment.* Ottawa: WDL Publications.

Babbie, E. (1990). *Survey Research Methods.* Belmont, Calif.: Wadsworth.

Baird, G., Gray, J., Isaacs, N., Kernohan, D., and McIndoe, G. (1996). *Building Evaluation Techniques,* Wellington, New Zealand: McGraw-Hill.

Bainbridge, W. (1999). Cyberspace: Sociology's natural domain. *Contemporary Sociology* 28(6):664-667.

Bakens, W. (2001). Thematic Network PeBBu—Performance Based Building—Revised Workplan, Rotterdam: CIB.

Balaram, S. (2001). Universal design and the majority world. In: Preiser, W.F.E., and Ostroff, E. (Eds.) *Universal Design Handbook.* New York: McGraw-Hill.

Basi, R. (1999). WWW response rates to socio-demographic items. *Journal of the Market Research Society* 41 (4):397-401.

Becker, F., and Steele, F. (1995). *Workplace by Design: Mapping the High Performance Workscape.* San Francisco, Calif.: Jossey-Bass.

Bordass, W., and Leaman, A. (1997). Future buildings and their services: Strategic considerations for designers and clients. *Building Research and Information* 25(4):190-195.

Bourdeau, Luc (Ed.) (1999). *Agenda 21 on Sustainable Construction,* CIB Report Publication 237. Rotterdam, Holland.

Bradley, N. (1999). Sampling for Internet surveys: An examination of respondent selection for Internet research. *Journal of the Market Research Society* 41(4):387-395.

Brager, G., Heerwagen, J., Bauman, F., Huizenga, C., Powell, K., Ruland, A., and Ring, E. (2000). *Team Spaces and Collaboration: Links to the Environment.* Berkeley: University of California, Center for the Built Environment.

Brand, S. (1994). *How Buildings Learn, What Happens After They're Built.* New York: Penguin Books.

Brill, M., and Weidemann, S. (1999). Workshop presented at the Alt.Office99 Conference. San Francisco, Calif.

Brill, M., Margulis, S.M., and Konar, E. (1985). *Using Office Design to Increase Productivity* (2 vols.). Buffalo, N.Y.: BOSTI and Westinghouse Furniture Systems.

Burns, P. (Ed.). *SAM—Strategic Asset Management Newsletter,* Salisbury, South Australia: AMQ International.

Business Week (1996). The new workplace. April 29, pp.107-117.

Cameron, I., and Duckworth, S. (1995). *Decision Support.* Industrial Development Research Foundation.

Campbell, D.T. (1999). *Social Experimentation.* Sage Publications.

Center for Environmental Design Research (Center for the Built Environment). (1996). *Vital Signs.* Berkeley: University of California. www.cbe.berkeley.edu.

Center for Universal Design. (1997). *The Principles of Universal Design (Version 2.0).* Raleigh, N.C.: North Carolina State University.

Centre scientifique et technique du bâtiment (1990). *Améliorer l'architecture et la vie quotidienne dans les bâtiments publics* Paris: *Plan construction et architecture.* Ministère des équipements, du logement, des transports et de l'espace.

CIB (1996). *A Model Post-Construction Liability and Insurance System,* CIB Publication 192. Prepared under the supervision of CIB W087.

Cohen, R., Bordass, W., and Leaman, A. (1996). *Probe: A Method of Investigation.* Harrogate, U.K.: CIBSE-ASHRAE Joint National Conference.

Cooper, C. (1973). *Comparison Between Architects' Intentions and Residents' Reactions, Saint Francis Place San Francisco.* Berkeley, Calif.: Center for Environmental Structure.

Consumer Reports. (2000). *Rating the Raters,* August 31.

Corry, S. (2001). Post-occupancy evaluation and universal design. In: Preiser, W.F.E., and Ostroff, E. (Eds.) *Universal Design Handbook.* New York: McGraw-Hill.

Cotts, D.G., and Lee, M. (1992). *The Facility Management Handbook.* New York: AMACOM, a division of American Management Association.

Davis, G., and Altman, I. (1996) *Territories at the Work-Place: Theory into Design Guidelines, in Man-Environment Systems,* Volume 6-1, 46-53. Also published, with minor changes, in Korosec-Serfati, P. (Ed.)

Appropriation of Space, Proceedings of the Third International Architectural Psychology Conference at Louis Pasteur University, Strasbourg, France, 1977.

Davis, G. et al. (1993). *Serviceability Tools Manuals, Volumes 1 and 2*. Ottawa: International Centre for Facilities.

Davis, G. et al. (2001) *Serviceability Tools, Volume 3—Portfolio and Asset Management: Scales for Setting Requirements and for Rating the Condition and Forecast of Service Life of a Facility—Repair and Alteration (R&A) Projects*. Ottawa: International Centre for Facilities.

Davis, G., and Szigetti, F. (1996). Serviceability tools and methods (STM): Matching occupant requirements and facilities. In Baird, G., Gray, J., Isaacs, N., Kernohan, D., McIndoe, G. (Eds.) *Building Evaluation Techniques*. New York: McGraw-Hill.

Dilani, A. (ed.) (2000). *Proceedings of the 3rd International Conference on Health and Design*. Stockholm: University of Stockholm.

Dillon, R., and Vischer, J. (1988). *The Building-in-Use Assessment Methodology* (2 volumes). Ottawa: Public Works Canada.

Eley, J., and Marmot, A.F. (1995) *Understanding Offices, What Every Manager Needs to Know About Office Buildings*. Middlesex, U.K.: Penguin Books.

Ellis, W.R., and Cuff, D. (1989). *Architects' People*. New York: Oxford University.

Farbstein, J., and Kantrowitz, M. (1989) Post-occupancy evaluation and organizational development: The experience of the United States Postal Service. In Preiser, W. (Ed.) *Building Evaluation*. New York: Plenum Press.

Federal Facilities Council. (1998). *Government/Industry Forum on Capital Facilities and Core Competencies*. Washington, D.C.: National Academy Press.

Fisk, W., and Rosenfeld, A.H. (1997). Estimates of improved productivity and health from better indoor environments. *Indoor Air* 7:158-172.

Flagg, G. (1999). Study finds major flaws in San Francisco main library. *American Libraries* 30(9):16.

Friedmann, A., Zimring, C., and Zube, E. (1978). *Environmental Design Evaluation*. New York: Plenum Press.

General Services Administration, Office of Governmentwide Policy, Office of Real Property. (1997). *Office Space Use Review*. Washington, D.C.

General Services Administration, Office of Governmentwide Policy, Office of Real Property. (1998). *Government-wide Real Property Performance Results—Government-wide baseline—Private Sector Performance—Building Profiles*. Washington, D.C.

General Services Administration, Office of Governmentwide Policy, Office of Real Property. (1998). *Government-wide Real Property Performance Measurement Study*. Washington, D.C.

General Services Administration. (1999). *The Integrated Workplace: A Comprehensive Approach to Developing Workspace*. Washington, D.C.: Office of Governmentwide Policy and Office of Real Property.

Gibson, E.J. (1982). *Working with the Performance Approach in Building*. CIB Report, Publication 64. Rotterdam, Holland.

Glover, M. (Ed.) (1976). *Alternative Processes; Building Procurement, Design and Construction*. Occasional Paper Number 2. Montreal, Quebec: Industrialization Forum.

Grantham, C. (2000). *The Future of Work: The Promise of the New Digital Work Society*. New York: McGraw-Hill, Commerce Net Press.

Gray, J. (in press) *Innovative, Affordable, and Sustainable Housing*. Proceedings of the CIB 2001 Triennial Congress. Rotterdam, Holland.

Gregerson, J. (1997). Fee not-so-simple. *Building Design and Construction*. August 30-32.

Guimaraes, M.P. (2001). Universal design evaluation in Brazil: Developing rating scales. In: Preiser, W.F.E., and Ostroff, E. (Eds.) *Universal Design Handbook*. New York: McGraw-Hill.

Hargreaves, D. (2001). Junk e-mail costs online surfers $14 billion a year: EU report. *Financial Post*. February 3, p. D9.

Hattis, D.B., and Ware, T.E. (1971). *The PBS Performance Specification for Office Buildings*. Prepared for the Office of Construction Management, Public Buildings Service, General Services Administration, by the Building Research Division, Institute for Applied Technology, National Bureau of Standards, U.S. Department of Commerce. Washington, D.C.: NBS Report 10 527.

Heerwagen, J. (2000). Green buildings, organizational success and occupant productivity. *Building Research and Information*. 28 (5/6):353-367.

Horgen, T.H., Joroff, M.L., Porter, W.L., and Schon, D.A. 1999. *Excellence by Design: Transforming Workplace and Work Practice*. New York: Wiley.

Huber, G.P. (1991). Organizational learning: The contributing processes and the literature. *Organization Science* 2:88-115.

International Organization for Standardization (1994). *ISO 9000 Compendium—International Standards for Quality Management, 4th Edition*. Geneva, Switzerland.

International Organization for Standardization 9000 (in process of re-edition)—Guidelines 9001 and 9004.

Joiner, D., and Ellis, P. (1989). Making POE work in an organization. In: Preiser, W. (Ed.) *Building Evaluation*. New York: Plenum Press.

Jones, L. (2001). Infusing universal design into the interior design curriculum. In: Preiser, W.F.E., and Ostroff, E. (Eds.) *Universal Design Handbook*. New York: McGraw-Hill.

Joroff, M., Louargand, M., Lambert, S., and Becker, F. (1993). *Strategic Management of the Fifth Resource: Corporate Real Estate*. Industrial Development Research Foundation.

Kantrowitz, M., and Farbstein, J. (1996). POE delivers for the post office. In: Baird, G., Gray, J., Isaacs, N., Kernohan, D., McIndoe, G. (Eds.) *Building Evaluation Techniques*. New York: McGraw-Hill.

Kaplan, R.S., and Norton, D.P. (1996). *The Balanced Scorecard*. Boston: Harvard Business School Press.

Kaye, B., and Johnson, T. (1999). Research methodology: Taming the cyber frontier. *Social Science Computer Review* 17(3):323-337.

Kernohan, D., Gray, J., and Daish, J. (1992). *User Participation in Building Design and Management : A Generic Approach to Building Evaluation*. Oxford: Butterworth Architecture.

King, N. (2000). What are they thinking? *PC Magazine (*February 8): 163-178.

Lorch, R. (Ed.) (2001). Post-occupancy evaluation (Special Issue) *Building Research & Information* 29(2).

Knocke, J. (1996). *A Model Post-Construction Liability and Insurance System*. CIB Publication 192, prepared under the supervision of CIB W087. Rotterdam, Holland.

Leaman, A., Cohen, R. and Jackman, P. (1995). Ventilation of office buildings: Deciding the most appropriate system. *Heating and Air Conditioning* (7/8):16-18, 20, 22-24, 26-28.

Loftness, V., et al. (1996). *Re-valuing Buildings: Invest Inside Buildings to Support Organizational and Technological Change Through Appropriate Spatial, Environmental and Technical Infrastructures*. Steelcase, Inc.

Lundin, B.L. (1996) Point of departure—Standard offers a new way to determine if a building measures up. *Building Operating Management* (September):154.

Lynn, M., and Davis, G. (1998). *The Need to Adopt an Auditable Commercial Real Estate Building Measurement Standard for Purposes of Securities Disclosure (A Ticking Time Bomb)*. International Center for Facilities.

Markus, L. (1990). Toward a "critical mass" theory of interactive media. In Fulk, J., and C. Steinfield (Eds.) *Organizations and Communication Technology*. Newbury Park, Calif.: Sage. pp. 194-218.

Maslow, H. (1948). A theory of motivation. *Psychological Review* 50:370-398.

McGregor, W., and Then, D.S. (1996). *Facilities Management and the Business of Space*. Arnold, a member of the Hodder Headline Group.

McLaughlin, H. (1997). Post-occupancy evaluations: "They show us what works and what doesn't." *Architectural Record* 14.

Mill, P., and Kaplan, A. (1982). *A Generic Methodology for Thermographic Diagnosis of Building Enclosures.* Ottawa: Public Works Canada (now PWGS: Public Works and Government Services Canada).

Mohr, A. (2000). The new metrics—going beyond costs per square foot to measure building performance. *Today's Facility Manager* (February).

Morgan Stanley Dean Witter. (2001). *Infoworld* March 12, p.16.

Nasar, J.L. (Ed.) (1988). *Environmental Aesthetics: Theory, Methods and Applications.* Cambridge, Mass.: MIT Press.

National Research Council. (1987). *Post-Occupancy Evaluation Practices in the Building Process: Opportunities for Improvement.* Washington, D.C.: National Academy Press.

Neufeld, E. (1997). Where are the audiences going? *MediaWeek* May 5, pp. S22-S29.

Norman, D. (1993). *Things That Make Us Smart: Defending Human Attributes in the Age of the Machine.* Reading, Mass.: Addison-Wesley.

Nua Internet Surveys. (2000). http://www.nua.ie/surveys/how_many_online.

O'Mara, M. (1999). *Strategy and Place: Managing the Corporate Real Estate in a Virtual World.* The Free Press.

O'Mara, M. (Ed.) (2000). Special issue on portfolio management. *Journal of Corporate Real Estate* 2(2).

Ostroff, E. (1997). Mining our natural resources: The user as expert. *Innovation* 16(1).

Ouye, J.O. (1998). Measuring workplace performance: Or, yes, Virginia, you *can* measure workplace performance. Paper presented at the AIA Conference on Highly Effective Facilities, Cincinnati, Ohio: March 12-14.

Ouye, J.O., and Bellas, J. (1999). *The Competitive Workplace.* Tokyo: Kokuyo (fully translated in English and Japanese).

Pedersen, A. (2001). Designing cultural futures at the University of Western Australia. In: Preiser, W.F.E., and Ostroff, E. (Eds) *Universal Design Handbook.* New York: McGraw-Hill.

Perseus. (2000). Survey 101—A complete guide to a successful survey. www.perseus_101b.htm.

Pike, P. (2001). Technology rant: I'm tired of feeling incompetent! *PikeNet* February 11. www.pikenet.com.

Preiser, W.F.E. (1983). The habitability framework: A conceptual approach toward linking human behavior and physical environment. *Design Studies* 4(2).

Preiser, W.F.E., Rabinowitz, H.Z., and White, E.T. (1988). *Post-Occupancy Evaluation.* New York: Van Nostrand Reinhold.

Preiser, W.F.E. (1991). Design intervention and the challenge of change. In: Preiser, W.F.E., Vischer, J.C., and White, E.T. (Eds.) *Design Intervention: Toward a More Humane Architecture.* New York: Van Nostrand Reinhold.

Preiser, W.F.E. (1996). POE training workshop and prototype testing at the Kaiser-Permanente medical office building in Mission Viejo, California, USA. In Baird, G., et al. (Eds.) *Building Evaluation Techniques.* London: McGraw-Hill.

Preiser, W.F.E., and Stroppel, D.R. (1996). Evaluation, reprogramming and re-design of redundant space for Children's Hospital in Cincinnati. *Proceedings of the Euro FM/IFMA Conference*, Barcelona, May 5-7.

Preiser, W.F.E. (1997). Hospital activation: Towards a process model. *Facilities* (12/13):306-315.

Preiser, W.F.E., and Schramm, U. (1997). Building performance evaluation. In: Watson, D., et al. (Eds.) *Time-Saver Standards: Architectural Design Data.* New York: McGraw-Hill.

Preiser, W.F.E. (1998). *Health Center Post-Occupancy Evaluation: Toward Community-Wide Quality Standards.* Proceedings of the NUTAU/USP Conference, Sao Paulo, Brazil.

Preiser, W.F.E. (1999). Post-occupancy evaluation: Conceptual basis, benefits and uses. In: Stein, J.M., and Spreckelmeyer, K.F. (Eds.) *Classical Readings in Architecture.* New York: McGraw-Hill.

Preiser, W.F.E., and Schramm, U. (2001). Intelligent office building performance evaluation in the cross-cultural context: A methodological outline. *Intelligent Building* I(1).

Raw, G. (1995). *A Questionnaire for Studies of Sick Building Syndrome.* London.: BRE Report. Construction Research Communications.

Raw, G. (2001). Assessing occupant reaction to indoor air quality. In Spengler, J., Samet, J., McCarthey, J. (Eds.) *Indoor Air Quality Handbook.* New York: McGraw-Hill.

Ripley Architects. (2000). *San Francisco Public Library Post Occupancy Evaluation Final Report.* San Francisco: Ripley Architects.

Rondeau, E.P., Brown, R.K., and Lapides, P.D. (1995) *Facility Management.* New York: John Wiley & Sons.

Schaefer, D., and Dillman, D. (1998). Development of a standard e-mail methodology. *Public Opinion Quarterly* 62:378-397.

Schein, E.H. (1995). *Learning Consortia: How to Create Parallel Learning Systems for Organization Sets* (working paper). Cambridge, Mass.: Society for Organizational Learning.

Schiller, G. et al. (1988). *Thermal Environments and Comfort in Office Buildings.* CEDR-02-89. Berkeley: University of California, Center for Environmental Design Research.

Schiller, G., et al. (1989). Thermal comfort in office buildings. *ASHRAE Journal.* 26-32

Schneekloth, L.H., and Shibley, R.G. (1995). *Placemaking: The Art and Practice of Building Communities.* New York: Wiley.

Sheehan, K., and McMillan, S. (1999). Response variation in e-mail surveys: An exploration. *Journal of Advertising Research* 39(4):45-54.

Shibley, R. (1982). Building evaluations services. *Progressive Architecture* 63(12): 64-67.

Shibley, R. (1985). Building evaluation in the main stream. *Environment and Behaviour* (1):7-24.

Sims, W.R., Joroff, M., and Becker, F. (1998). *Teamspace Strategies: Creating and Managing Environments to Support High Performance Teamwork.* Atlanta: IDRC Foundation.

Smith, P., and Kearny, L. (1994). *Creating Workplaces Where People Can Think.* San Francisco: Jossey-Bass.

Spillinger, R.S., in conjunction with the FFC (Federal Facilities Council). 2000. *Adding Value to the Facility Acquisition Process: Best Practices for Reviewing Facility Designs.* Washington, D.C.: National Academy Press.

Steele, F. (1973). *Physical Settings and Organizational Development.* Reading, Mass.: Addison-Wesley Publishing Company.

Steele, F. (1986). *Making and Managing High-Quality Workplaces.* New York:Teacher College Press, Columbia University.

Stewart, T.A. (1999). *Intellectual Capital.* New York: Doubleday

Sullivan, E. (1994). Point of Departure—ISO 9000: Global benchmark for manufacturers may be future facility challenge. *Building Operating Management* (October):96.

Swartz, P. (1991). *The Art of the Long View.* New York: Doubleday.

Swoboda, W., Muhlberger N., Weitkunat R., and Schneeweib, S. (1997). Internet surveys by direct mailing. *Social Science Computer Review* 15(3):242-255.

Taylor, H. (2000). Does Internet research work? *International Journal of Market Research* 42(1) 51-63.

Teicholz, E. (Ed.) (2001). *Facilities Management Handbook.* New York: MacGraw-Hill.

Thompson, H. (2000). *The Customer-Centered Enterprise—How IBM and Other World-class Companies Achieve Extraordinary Results by Putting Customers First.* New York:McGraw-Hill.

USA Today (2001). Searching the Web in native language. February 27, 7B.

U.S. Department of Commerce. (1999). Falling through the net: Defining the digital divide. National Telecommunications and Information Administration. http://www.ntia.doc.gov/ntiahome/fttn99.

Ventre, F. (1988). Sampling building performance. Paper presented at *Facilities 2000 Symposium.* Grand Rapids, Mich.

Vischer, J. (1985). The adaptation and control model of user needs in housing. *Journal of Environmental Psychology* 5:287-298.

Vischer, J. (1989). *Environmental Quality in Offices.* New York: Van

Nostrand Reinhold.

Vischer, J. (1993). Using occupancy feedback to monitor indoor air quality. *ASHRAE Transactions* 99(pt. 2).

Vischer, J. (1996). *Workspace Strategies: Environment as a Tool for Work.* New York: Chapman and Hall.

Vitruvius. (1960). *The Ten Books on Architecture* (translated by M.H. Morgan). New York: Dover Publications.

von Foerster, H. (1985). *Epistemology and Cybernetics: Review and Preview.* Milan, Italy: Casa della Cultura.

Watson, C. (1996). *Evolving design for changing values and ways of life.* Paper presented at the IAPS14, Stockholm.

Watson, D., Crosbie, M.J., and Callender, J.H. (Eds.) (1997). *Time-Saver Standards: Architectural Design Data.* New York: McGraw-Hill (7th Edition).

Welch., P. (Ed.) (1995). *Strategies for Teaching Universal Design.* Boston, Mass.: Adaptive Environments Center.

Welch, P., and Jones, S. (2001). Teaching universal design in the U.S. In: Preiser, W.F.E., and Ostroff, E. (Eds.) *Universal Design Handbook.* New York: McGraw-Hill.

Wineman, J. D. (Ed.) (1986). *Behavioral Issues in Office Design.* New York: Van Nostran Reinhold.

Zeisel, J. (1975). *Inquiry by Design.* New York: Brooks-Cole.

Zeisel, J. (1975). *Sociology and Architectural Design.* New York: Russell Sage Foundation.

Zeisel, J. (in press). *Inquiry by Design,* 2nd edition. New York: Cambridge University Press.

Zimring, C.M., and Reizenstein, J.E. (1981). A primer on post-occupancy evaluation. *Architecture* (AIA Journal) 70(13):52-59.